MARKUP & PROFIT

A CONTRACTOR'S GUIDE

Michael C. Stone

Craftsman Book Company
6058 Corte del Cedro / P.O. Box 6500 / Carlsbad, CA 92018

Acknowledgments

The author wishes to pass along a special thanks to a few who have given so much, both to the industry, and to the development and writing of this book.

- To Devon, my wife. She is the perfect wife and mother. I hope each of you has a helpmate as good for you as Devon is for me.

- To Laurence Jacobs, Genie Runyon and the rest of the gang at Craftsman Book Company, who have prodded, yelled, passed out "awards" and coerced this book through from start to finish. One would be hard-pressed to find a better bunch of folks to work with on any endeavor.

- To my friends in the business, Bill Carr of Malvern, Pennsylvania; Tom Dwyer of Dunwoody, Georgia; Eugene Peterson of Salt Lake City; and Clai Porter of Anchorage, Alaska. Whenever I've needed something, knowledge, advice, or help in any way, they've been there. My thanks to each of you.

I would like to dedicate this book to the memory of René Hamel of West Covina, California
René, who was involved in residential remodeling, was an excellent businessman and one of the best friends one could hope to have.

Looking for other construction reference manuals?
Craftsman has the books to fill your needs. **Call toll-free 1-800-829-8123**
or write to Craftsman Book Company, P.O. Box 6500, Carlsbad, CA 92018 for
a **FREE CATALOG** of over 100 books, including how-to manuals,
annual cost books, and estimating software.
Visit our Web site: http://www.craftsman-book.com

Library of Congress Cataloging-in-Publication Data

Stone, Michael C.
    Markup & profit : a contractor's guide / by Michael C. Stone.
        p.   c.m.
    Includes index.
    ISBN 1-57218-071-4
    1. Building--Estimates. 2. Contractors' operations--Costs.
3. Construction contracts.   I. Title.   II. Title: Markup and profit.
TH435.S825   1998
690'.0682--dc21   98-42265
                                   CIP

©1999 Craftsman Book Company
Third printing 2001

Contents

1 Looking at Markup in the Construction Business — 5

So You Want to Be a Successful Contractor ..6
And If You're Already a Contractor? 11
And the Formula Is … 18

2 Understanding Markup — 19

Symptoms of Impending Failure 19
Three Major Causes of Business Failure 23
The Terminology of the Industry 26
Determining Your Financial Requirements ..27

3 Establishing the Correct Markup for Your Company — 53

The Basic Formula for Markup..............53
Let's Do Some Sample Problems 54
Sliding Scale Markup 67
A Review of Volume, Overhead and Markup .69
Problems to Watch Out For 81
Job Supervision and Markup 85

4 Sell Your Services — at a Profit — 89

The Basics of Attracting Sales 89
Your Telephone Is Ringing, Now What? 96
The Basic Steps of the Sale 100
Some Final Notes About Sales 115

5 Writing Contracts — 121

The Importance of a Detailed Contract 122
Writing Your Contract Documents 126
Good Contracts Have
 Well-Defined Pay Schedules 137

6 Change Work Orders, Other Forms and Your Markup — 149

Using Change Order Forms 149
Other Forms 160

7 The Mathematics of Your Business — 169

Setting and Keeping a Budget 169
Margins vs. Markup 178
Break Even 180
Math, Formulas and Ratios
 for Your Company 183
Computers 189

8 Bean Counters and Your Business — 197

Dealing with the Bean-Counter Mentality ...197
Financing with Two
 Bankers and One Broker 199
Working for Bankers 202
Insurance Companies, Their
 Pay Schedule and Your Markup 203

9 Justifying Your Markup — 207

Communication 207
Classes, Seminars and Conventions 211

10 Employees and Your Markup — 213

Training 213
Productivity 214
Schooling on Markup 217

11 Listening to the Experts — 219

Recognizing an Expert 219
Increasing Your Expertise in Your Field 221
Hiring Consultants 223
Some Good Advice 224
In Conclusion, May I Say Thank You ... .228

Problems to Solve — 229

Problem 1: Getting Started 230
Problem 2: A Fourth-Year Business 232
Problem 3: Correcting a Disaster 236
Problem 4: All That Glitters Is Not Gold!!! 242
Problem 5: Evaluating a Quarterly Review ...247
Problem 6: Boom & Bang Construction 251
Congratulations! 263

Appendix

Blank Forms 265
Educational and Construction Resources 305

Index 312

Chapter 1
Looking at Markup in the Construction Business

For years I've looked for a good book on how to establish markup for a construction company, or on how to deal with problems I've had with markup. The books I needed didn't exist, so I decided to fill that void. I wrote this book simply because this information is needed. Yes, you may be able to find an occasional article in a trade magazine that covers markup and its associated problems, but you'll never find one that goes into any depth or covers the theory that I'll cover for you here.

I chose the title of this book with care. When you've finished reading it, I guarantee that you'll thoroughly understand what markup is and how to arrive at it. And more importantly, you'll know how and why you should apply it to your job costs to arrive at the right sales price for your work. The information in this book can all be backed up with simple mathematical principles or formulas. Although some of the material may seem like it's nothing more than my opinion, believe me, it isn't. I haven't been standing in a college classroom lecturing about something I've never done — I'm a mud-on-my-boots contractor. I've probably crawled further under a house than most contractors have traveled away from home. I have, along with a number of other very good contractors, paid the high tuition fees at the School of Hard Knocks to learn the information that you'll find in this book.

Construction contracting as a business has the highest failure rate of any business in America today. According to Dun & Bradstreet, the U.S. Dept. of Commerce and all the available statistics, the odds for success in this business are stacked high against you. If you're under the delusion that statistics only apply to the other people in this business . . . you'd better keep reading. I wrote this book specifically to enlighten those of you with that attitude.

Does this mean you're bound to fail? No! You don't have to be a victim of this high failure rate. In fact, you can avoid it altogether if you have the discipline to learn a few commonsense things that the professionals in the construction business do on a daily basis.

I used to feel sorry for contractors whose businesses failed. But I don't feel that way anymore, and I'll tell you why. There are statistics readily available that show clearly why those contractors failed or went bankrupt. The facts, figures, and information about business failures are there for anyone willing to invest a little time and effort to evaluate it.

In this book I'm going to tell you how *successful* contractors work. You'll hear their stories: we'll discuss their triumphs and their occasional failures. You may be surprised by some of the clever approaches these contractors have taken to solve the same problems that you've faced. You'll learn their methods for success, and when you finish this, you should be able to apply those methods to aid in your own success.

Speaking of learning, you'll find that continuing education is one of the key elements that all successful companies have in common, whether they're contracting businesses or other businesses. So give yourself a pat on the back for taking the initiative to better yourself by reading this book. You've already crossed the biggest hurdle of any learning process — getting started. Statistically, if you're already a contractor, you're one of approximately 4 to 5 percent of all contractors that work at improving their companies through an ongoing process of education. We'll talk some more about that later on. Right now, let's begin that education.

So You Want to Be a Successful Contractor

Some of you may not be contractors yet, but you think you'd like to be. That's admirable. It shows energy, initiative and hope of self-determination — all very good qualities. You want to get out of the daily grind of working for other people, and direct a good company that does lots of business and puts the profits in your pocket. You want to make some *good* money!

The goal (the fancy phrase would be "the mission statement") of this book is to review the basics of what it takes to survive in the residential construction business. More specifically (and more importantly), you'll learn how to think your problems through, get paid, *and* make a profit. You can sell all the jobs available in your area, remodeling or new home, residential or commercial, *but if you don't get paid adequately when the jobs are completed, your efforts are wasted.* Making a profit is what this book is all about.

I've gathered the information, problems and solutions used in this book from people in the same business situation that you're in. They get up every morning to face the same problems that you do. I've made about every mistake anyone can make in this business — and several more than once. Unlike many of the so-called "industry experts" that are quick to tell you how to run your business, you'll find that the help offered here is both practical and timely. It works in today's business world.

In 1988 I had the opportunity to talk with Dave Sauer, the former owner and CEO of *Qualified Remodeler* Magazine. During our conversation he said, " . . . the American public today is getting the biggest bargain in history when they contract with most contractors to have work done on their homes. Contractors simply do not charge enough for what they do." Those words are just as true today as they were then. And, contrary to what many people think, this applies to commercial construction as well as residential construction and remodeling.

If you read this book cover to cover, *and work the problems*, when you finish you'll know what you should be charging for your work. Additionally, you'll have had the benefit of my advice as well as that of a number of other successful contractors on what it takes to survive in this business. The construction business is very demanding, but it can also be very rewarding. Do you have what it takes? Let's take a look.

Five Basics for Survival

In my opinion, there are five basic criteria that you should meet before you venture into your own construction business. This applies to anyone who wants to build new homes or commercial buildings, do residential or commercial remodeling, or develop a specialty contracting business like electrical, plumbing, drywall or roofing. If you don't meet the criteria outlined, then you'd better have a very compelling reason to pursue a goal as difficult as starting your own construction company. As Michael Gerber says in his great book *The E-Myth* (the next book you should read), there's much more to being in business than simply being a good technician.

If I were going to lend you the money to start a new construction business, here are the business basics that I would expect you to have:

1. A journeyman's level of competency
2. An understanding of the sales process
3. A commitment to education
4. An understanding of markup
5. A readiness and willingness to work hard

A Journeyman's Level of Competency

Although not mandatory, you should have at least a journeyman's level of competency in one or more building trades. That means at least four solid years working at a particular trade with people who have been well trained and know what they're doing. As a company owner and employer, you should be good enough at what you do to teach others. You want them to be able to rise to, or exceed, your skill level. Part of being in business is being a good teacher. But if you're not competent at what you do, no amount of teaching skill will overcome that deficiency.

What about those who attempt to get into construction without a trade background? While it's true we get many construction workers who are also teachers, firefighters, janitors, college students, lifeguards, and from other professions that don't require them to be on the job 9 to 5, few of them survive if they attempt to become contractors. They may have strong backs and good business management skills, but they don't have the necessary practical work experience behind them to succeed as contractors. So if you're among those ranks, review your assets carefully. It'll take a ton of them to make up for your deficiency in good trade skills.

An Understanding of the Sales Process

"Nothing happens until somebody sells something." When you start your own company, you're in sales whether you like it or not. You need to have a good understanding of the sales process. I've taught sales to literally hundreds of people involved in the sale of construction services, either in new homes or remodeling. Working with and watching those people has shown me that it takes at least five years of full-time selling to really understand the sales process.

Zig Ziglar, noted sales and motivational trainer, says that selling is the highest-paying hard work and the lowest-paying easy work that you can do. If you start a company of any kind, you have lots of sales work ahead of you. In order to make money, you must be able to sell your product. When you combine the energy that you need for sales with the work involved in starting and/or running a business, you have your plate full!

A Commitment to Education

You should have an ongoing commitment to education. Knowledge, like the pursuit of perfection, is a lifetime endeavor. To be a successful contractor, you must be a student of both construction and business. If you haven't read at least one book in the past 30 days (not counting this one) on a subject related to your particular trade or business in general, you probably haven't yet made this commitment.

An Understanding of Markup

You should have a thorough understanding of good accounting procedures and how they apply to your company. Even more important, you need to really understand how to establish and maintain the correct markup for your company. If you don't know the fundamentals of how to arrive at the right sales price for your work, you probably won't survive in the construction business.

It's now time for the first of many gut checks that we'll do in this book. How do you establish your markup? ***Right now, in one simple sentence, write down the correct formula that you think you should use to establish your markup.*** Let's see if you know as much about markup as you think.

Even if you're not sure of the correct formula, write down your best guess. After you've written it down, set your formula aside. We'll check it a bit later. Few contractors know the best way to establish their markup. But if you're not using the right method, don't worry — at least not yet. You're reading this book to learn how, and that's good. By the time you finish, you'll know the subject inside and out.

A Readiness and Willingness to Work Hard

The final building block to making your business a success is simple hard work. You'd better be ready and willing to put in at least 60 hours a week for the first three to five years. That's right: *60 hours*, minimum. Of course, you might be good enough that you do everything right the first time. Then you can cruise through 40 hour weeks. But don't count on it. In my 30-plus years in this business, I've yet to meet anyone who's started and developed a successful company by working less than 60 hours a week. You think having your own business means you sit back and watch the money come rolling in while your employees do all the work? Dream on!

Honesty and Attitude

Those are the five basics for a successful start in the construction business. But there's more to it when you're working with people in a service capacity. You also have to have some good people skills. So if you add honesty and the right attitude towards people to the list, you're ready to make a good beginning.

Honesty

In contracting, as in other businesses, there are always a small percentage of people who don't do business with candor and honesty. We read or hear about home improvement scams on a regular basis, and you can drive through new subdivisions almost anywhere in the U.S. and find owners suing builders because what was "promised" wasn't what was delivered.

Briefly and to the point: To be successful, you must conduct yourself and your business in a manner beyond reproach at all times. For peace of mind, for the acceptance of the people you work with and the people in the community that you work in, there's simply no other way to do business. You've heard the old saying, "You can fool some of the people some of the time . . ." and so on. It's true. I've never, in all my years in the business, found any individual or company to be successful if they were anything less than completely honest.

This is the second gut check. Only you know if you're using a completely honest approach to your business and your life. As Zig Ziglar says, "honesty is what you do in the dark." Make sure your business practices can stand up to the light of day.

An Example of the Attitude of Success

Now I'll give you an example of the "attitude" that I think you need to make it in the business world. This is a very short story about a beautiful little lady who owns a bakery in the seaside community of Lincoln City, Oregon.

Recently, my wife and I took our two daughters to the coast for the weekend. I have this belief that every good contractor, husband, father and lover should have a sweet roll on Saturday mornings. It gets you going, keeps you sharp, freshens your breath and helps put you on top of things. On this particular Saturday morning, I was up at 6:00 a.m. so I could read a bit as I do most every morning, and then I jumped in the car and drove down to the local bakery. I arrived at the bakery at 6:45 a.m. (please note the time) and was greeted by a charming lady with a radiant smile. "Good morning! How can we help you?" She obviously wanted to help me get my day started right. I paid for my rolls, received a nice "Thank you!" and went back to our motel room and the eager appetites of my family. We had a fine day, with walks on the beach, shopping at all the junk and antique shops in town, a short nap, a nice run on the beach, and dinner out!

 "Education is one thing that successful contractors have in common."

We drove by the bakery on our way back to the motel after dinner. I decided to stop to see if they'd be open on Sunday morning. The sign said they were open 6:00 a.m. to 6:00 p.m., Sunday through Saturday. Without my knocking or in any way trying to attract attention to myself, the woman who had waited on me that morning noticed me and came to the door. Even though it was 7:15 p.m. and long after closing, she opened the door and asked if she could help me. I told her I was just checking the store hours, and that I'd be back the next morning to get more sweet rolls. She invited me in. "Come on in and get your rolls now. It'll save you a trip here in the morning." She'd probably been there since before 6:00 a.m. that morning and it was after 7:00 p.m., but she still had the same smile and the same beautiful attitude. That's what it takes to be a success! This is a business owner with the correct mindset: she loves her business and enthusiastically serves her customers without even thinking about it. There's no way that she or her business can fail.

The question is, **can you run your business and serve your customers in the same way?** This is another gut check. By the way, a gut check is a kind of yardstick that you can measure your own work and your own company against. I'm simply giving you an opportunity to see where you are in the general scheme of things. So how do you measure up? You're the only one who'll ever know if the measurement is accurate and acceptable. If you're 100 percent honest with yourself, you already know the answer.

Do You Have a Great Future?

If you don't measure up to these standards, what are you going to do about it? The first thing is to finish reading this book. As you read, map out a course of action to learn the things you don't know so that you'll meet those basic standards. Then *follow up* and take that action. Finally, when you're really ready, get your license and get to work. You'll have a great and profitable future ahead of you.

If, on the other hand, you insist on getting into the business without meeting the standards, be prepared to join the ranks of contractors that fail. You'll have plenty of company. Why? Because too many contractors go into business without the proper preparation and background. In a nutshell, they don't know enough to charge the right amount for the work that they do. Knowing how to build something is *less than half* the knowledge you'll need to do business and make a profit in construction. You'll do fine working for someone who does, but the odds are against you making it on your own.

Education

Education is one thing that successful contractors have in common. It's something they all pursue. And the older I get, the more I believe that it applies to all aspects of life, both personal and business.

There's more to education than sitting in a classroom at a school. You're in a classroom every day of your life. Unless you're independently wealthy, the grade you get in that classroom is the money you take home from your job. I've often had salespeople ask me how they'll know when they've become a "good" salesperson. The answer is very simple — you look at the W-2 you get at the end of each year. Your taxable income says it all! That's the number that counts. That's how good you are. You can apply that same measurement if you own and run your own business.

I believe it's almost impossible to read a book, or a magazine, or a newspaper article and not get at least one idea that you can use in your business. Not all the ideas are great, but if you find just one good idea that you can use, your time will have been well spent. We'll continue to talk about education as we go along. By the end of the book, I hope you're convinced of the importance of education.

And If You're Already a Contractor?

If everything were perfect in your world, you probably wouldn't have bought this book. So let's be honest here. You've reached the conclusion that you need some help, and made the decision to seek it out. Good for you! You may not find all your answers here, but you're on the right track.

Build on the experience of the contractors we'll discuss in this book to help you solve your problems.

Make no mistake, *everyone* has problems! So what's the difference between those who succeed and those who don't? It's simple. Successful contractors know how to think through a problem to get to the root cause, *and then they deal with it*. Those who become statistics try to deal with the result of the problem without effectively eliminating the cause. They get by, often thinking that if you ignore something long enough, it'll go away. And it will — along with their business.

Keep an Open Mind or "Attitude Adjustment Time"

The attitudes and dispositions of contractors are fairly typical. Would you find it surprising if I told you that most of the folks in contracting have strong egos, are independent, and want to be self-directed in their business? It's true, you know! Some of us might even be accused of being stubborn! This is neither good nor bad, it's just the way we are — and I'm just as guilty. This attitude of ours is the very thing that makes us want to have our own businesses. It gives us the *go power* that we need to make things happen. However, it can also work against us.

In any situation, you must constantly check to be sure that you're keeping an open mind, especially about any new information you're taking in. And this book is full of new ideas. All I ask is that you're open and receptive to something new. Give yourself and this book a chance.

How do you know if you're keeping an open mind? Here are few things to watch for:

- If you immediately start to debate my ideas in your mind, that's a sure sign that you don't want to let go of your thinking on that subject. Save your debates for later, and continue on through the book. Get the whole picture before you jump to any conclusions.

- If you find your arms folded, fists clenched, or your legs crossed tightly, again you may be resisting the new idea. Keep loose, unfold and unclench, go for a walk and then come back to it. Stay open.

- Last but not least, don't just read, get involved. Fill in the blanks, do the problems, compare the ideas outlined in this book with what you are doing within your own company.

Allow yourself some time to get used to new information. Give your subconscious mind a chance to work on the ideas. Some of this stuff takes a while to soak in. I've been in a state of soak for over 30 years. When you start feeling like a sponge, then you're getting there!

Review Your Company Policies

In the best-run companies, policies are analyzed and fine-tuned on a regular basis. Every well-run company has a Method of Operation Manual (M.O.M.) that clearly outlines how the company is to be managed. You may call your manual by some other name, but if you have one, it will dictate how your company operates. However, while working with contractors on business management problems over the last 18 years, it's been my observation that fewer than 6 percent of all construction company owners in the United States have taken the time to write a M.O.M. for their business.

Write a Method of Operation Manual

If you haven't completed a M.O.M. for your company, you'll find it's no small task. When you try to put your ideas in writing, you have to give them a lot of thought. Start thinking about how to compile your M.O.M. as you go through this book. As you come across new ideas, you can analyze them and decide if they should be covered in your manual.

Begin the process by making an outline of your company and how it operates. Then expand the outline to describe exactly how each part should function. Cover all areas of your business. Include your employees, your jobs, your customers and your service and supply companies, and how you interact with each of them. As you build your M.O.M., have the people in your company review it. Everyone should be able to give you some feedback, especially on sections that concern their particular job areas.

A manual written in this way will give you and your company direction and show you how to approach all aspects of your business. You're creating a goal for your company that you should make happen on a daily basis. It's like a road map that shows where you've been, where you are, and more importantly, where you're going. If it's done correctly, a stranger could take your manual and find the answer to any question about your company or how to handle any given situation. In short, they could run your company from your manual.

A M.O.M. is well worth the time and effort that it takes to put together. What kind of time are we talking about? If you're starting from scratch, it'll probably take you up to a year to write. If you're revising an existing M.O.M., plan on two to three months of work.

Set Goals

Only 4 to 6 percent of the contractors I've ever met have bothered to set goals, either for themselves or their company. There are all kinds of excuses for not putting your goals down in writing, and frankly, most people find them. If you've never set goals for yourself or your company, it's time to start. There are countless books, audiotapes and videotapes on how to do goal setting. Just like your M.O.M., goals are a road map of where you want to go.

Over 20 years ago, I decided to write down everything that I could think of that I might ever want to do in my life. I worked on this over a period of about six months, putting a lot of thought into it. I didn't, however, put any judgments or criteria on the items I listed; if I thought it might be interesting, fun, challenging, dangerous, or whatever, I wrote it down. Then I compiled all of the various lists into 186 goals of things I wanted to do with the rest of my life. Some years I've accomplished two or three items and checked them off the list. Some years I haven't completed any. But talk about fun! I even got a box of little gold stars to put by each one of goals that I completed. When every goal on a page is completed, I transfer the page to the "Goals Completed" section of my Goals book. Completing one of those goals gives me almost the same feeling as selling a remodeling job for $350,000 and knowing I'll make at least a 10 percent net profit on it no matter how the job turns out.

Set your goals, at least for your business. It will be the best investment of your time you'll ever make.

Start a Checklist for Making Money

You wouldn't think it's necessary to develop a checklist for making money, but it is. And you have to do it yourself. No one else can do it for you because you're the only one who knows what *your* motivation is. Your checklist should be a list of reminders of what you need to do to stay focused on your business — and you should review it each day. It's easy to fall into the trap of doing the things you like to do instead of doing the things that will make you money.

I read recently that one of the habits of successful people is to make a point of doing those things they *need* to do, whether they like to do them or not. That's what sets them apart. Other people avoid doing things that they don't like to do. Successful people don't like to do those things either. The difference is that *they do them anyway*. Your checklist will help keep you focused, and keep you doing all those things you need to do to make money.

Here are some examples of the items I suggest for your daily money-making checklist. Depending on your personality and what makes you tick, I suppose you can come up with five or six, maybe as many as nine or ten more items to add to your list.

- **I will accurately estimate each job and apply the correct markup for that job.**
  This statement keeps me focused as I sit down at my computer to compile a particular estimate for a given job. It helps me follow our company rule that any item on the estimate sheet that exceeds $300

must have a written quote from a specialty contractor or a supplier. If I do that, then apply the correct markup to the job costs, I'll arrive at a sales price that covers all our job costs and overhead expenses and gives the company at least an 8 percent net profit. I've estimated over 3,600 jobs, and even with that background I must guard daily against getting lazy and taking shortcuts.

- **I will focus on those things I can control, and ignore all other distractions.**
  This is some advice that I've heard from several very good contractors over the years. We've all spent a lot of years and money learning about staying focused.

- **I will focus on and do the most productive thing I can do at each moment.**
  This is another one that's used by many folks in the business. It's a favorite of Tom Hopkins, the noted sales trainer. He recommends it to all his disciples.

- **I will work a full eight-hour productive day today.**
  Keep track of your time on a half-hour basis for the next 30 days. I think you'll find that you waste a lot of your time on nonproductive activity. I tried it once. What an eye opener! I was spending more than half my waking hours on totally unproductive activities. Plain and simple, I wasn't staying focused on making money. I was doing all the fun things I liked to do, and not the things that put bread on the table. Try it yourself.

It might take a couple of months, or even longer, to arrive at a final checklist that's right for you. The end result, however, is that you'll start making more money. And that's what we're all about in this book — making money.

Fall in Love with Your Business!

One of the things that you must do to survive in this business is to "fall in love" with your business. I don't mean the remodeling you do or buildings that you build, but the business of being in business. Another way of saying the same thing would be "marry your business, not your work."

Too many people in construction get all caught up in the "quality" of their work. They want to build their business based on the "great work" they do. While this may be admirable, it's nonsense! Your objective should be to provide your customers with a good job, the job that you contract for at the quality that you've said you'd give them. In short, *you give them what they pay for.* That's a good-quality job — no more, no less. Don't be tempted to add that little piece of trim molding around the cabinets or full extension drawer guides to the kitchen drawers if they aren't already in the plans and budgeted for. Sure they'd be a nice addition to your kitchen remodel, but who's going to pay for them? You've got to keep your eye on the budget for the job. There's only so much money estimated for each job; when you know it'll cost more, you have to *stop*! It doesn't take too many of these little "extras" to eliminate any profit you may make on a job. If you have a good idea for a modest improvement and you take it to the owners *and they want to pay for it*, that's fine. If they don't want to pay for the change, leave it at the quality you've agreed on, and move on. When the job is complete, assemble the final job costs, analyze the job, and then make any adjustments to the way you'll estimate or build similar jobs in the future. Next time you may want to build those little improvements into your estimate.

Many company owners pick up on this approach quickly, while some employees don't. Watch your employees carefully to make sure they don't spend too much time on any phase of a particular job. They also tend to "fall in love" with the jobs they're building. That's why it's so important to give your job superintendent a list of the hours you estimated for each phase of the job before you start. They need to know where they should be at any given point of the job. That's how they know the time investment and the dollars available to complete that particular job phase correspond.

This tendency to spend too much time perfecting the job is one of the reasons that I subcontract out most of the work that's done by my own construction company. I have one employee (my job superintendent), and he knows to the penny and to the hour the dollars and time I've estimated for any given job. He checks all my estimates and we resolve any and all price issues before we give the quote to the customer. But first we get a firm written price quote for every job cost that exceeds $300. That's why almost every job we do comes in right on or below the budget.

If subs working on a job for us run over their quotes, that's their problem. I'm not being mean or malicious or unfair to our subs. We expect them to conduct their business in the same professional manner that we do. When they give us a written price quotation, we expect them to perform that work according to the specifications we gave them and at the price they gave us. We have a good working relationship with the subs we use. Normally, we only get one quote for each specialty on each job. We don't shop around, get three estimates, or any of the "we don't trust you" routine that so many general contractors waste time and effort on. We get a firm price quotation from people we trust, go with that number, and get on with the job at hand. That way, we can focus our efforts on getting the jobs done right and on budget.

On the other hand, some not-so-reliable contractors compile their estimates by the W.A.G. method. (If you don't know what that is, I'll explain it in the next chapter.) They figure they can always browbeat a sub down to recover any money that they lose because their estimate was too low. Or they can save money by using lower-quality materials than they sold the customer. "They'll never know the difference!" they say. I've even heard of contractors who've cut $3^1/_2$-inch insulation in half, doubling the amount of wall space that a roll of insulation will cover. That doesn't save a whole lot, but some of these guys figure "every little bit helps." As a last resort, they'll go back to the customer and ask for more money to complete their job. They might even threaten to pull off the job if the customer doesn't come up with the money. As you can well imagine, they don't get much repeat business.

You *can* get your jobs done well and on budget with your own employees. It just takes more planning and supervision. But don't try the excuse that you use your own employees instead of subs because that's the only way you can control the quality of your jobs. That just doesn't wash. If you pay a sub for work that's a lower quality than you'd expect from your own employees, whose fault is that? If you specified what you want on a particular job and the work isn't completed to your specifications or satisfaction, don't pay for the work until it's done right. The customer certainly won't!

"I have one employee (my job superintendent), and he knows to the penny and to the hour the dollars and time I've estimated for any given job. He checks all my estimates and we resolve any and all price issues before we give the quote to the customer."

What? You say you can't find a sub who'll do the work the way you want it done? Then keep looking for the right sub or rethink your expectations. Maybe you expect too much. Unless you're getting paid for perfection (and I'm sure you're not), relax your standards a bit and get the job done. Give them a good-quality job, keep the job moving, get it done and get out, and your customer will be happy. The longer it takes you to finish a job, and the more problems that arise, the less happy the customer will be with the result — that's human nature.

So, get excited about your business, fall in love with it, but not the work that you do or the people that you do it for. The daily mechanics of construction are too time-consuming for the owner of a construction company to get involved in. Hire good people and trust that they'll do a good job for you. Now that's not to say that you shouldn't work on any of the jobs. If you like the physical work of building, by all means, get right into it. But never forget that you have a business to run, and that takes top priority.

Here's a good rule of thumb that works for me if I'm tempted to get some hands-on involvement on one of my jobs. Let's say that I set my value to my company at $45 an hour. (That's not the real number, but a good one for this example.) If I can hire an individual or subcontractor to do that job for less than $45 an hour, I do it. If the cost of that particular job runs more than $45 an hour and I have the time, knowledge and ability to do that job, then I do it. And I smile all the way to the bank with the savings in my pocket. Easy enough!

And the Formula Is . . .

I asked you earlier in this chapter to write down a formula for establishing the markup for your company. Did you do it? If you did, you passed this gut check. If you're committed to success, you'll have your formula ready to compare to my formula.

Here's the formula for markup:

Markup = Total Volume Sold ÷ Job Costs

This formula isn't based on my personal opinion. It's a time-tested mathematical formula that works every time. You can check it with any CPA, and it'll pass the test every time. And did you notice that I didn't list any particular number to use for markup? There's a good reason for that. Every company must establish its own markup, using its own overhead and profit numbers. Using somebody else's numbers for your company isn't only foolish, it's suicidal.

To establish an effective markup, you need to:

1. Combine your job costs, your overhead and your profit into one final figure called Sales Price or Volume Sold.
2. Use that figure to come up with a markup that's right for your company.
3. Then actually use the markup number you arrive at.

This last area is where most contractors fail. They know, or should know, what they need to charge for their services, but they just don't do it. That's where I come in. I'll help you overcome any objections that you have to charging the amount you need to be successful in this business.

Read on.

Chapter 2

Understanding Markup

There's little debate that contractors involved in remodeling and new construction fail at an alarmingly high rate. In fact, there's solid evidence that well over 90 percent of all companies that obtain a business license as a general or specialty contractor will be out of business within ten years. If you've been around the industry a while, you can think back a few years and recall several companies that you thought were "doing well" but have since disappeared from the scene. See Figure 2-1.

So why, then, would anyone want to go into the construction business? Since 1980, I've been involved in the presentation of classes and seminars for people in the construction industry. We always ask "Why did you go into business for yourself?" And we get two answers more than any other. The first is that they want to be their own boss (self-determined). The second is that they want to make money based on their efforts, not just a wage or salary.

I can understand someone wanting to be their own boss and set their own direction. Normally, entrepreneurs are the types of people who want to control their own destiny. But what about this desire to put the profits from their own efforts directly into their own pockets? Well, making a profit is what being in business is all about, isn't it? That's the theory anyway, and that's what this book is about. But remember, profits are the result of a well-managed company that provides its customers with the product or service they want.

Symptoms of Impending Failure

Dun and Bradstreet has been keeping tabs on the cause of business failures since about 1920, and they list insufficient profits as the single largest cause of business failure. Put another way, the companies that failed had more bills left at the end of the month than they had money. Sound familiar?

Statistics, if applied with honesty and consistency, will always give you good dependable information on which you can base your business decisions. So why, if there's so much documented evidence on the high mortality rate,

Chapter 2: Understanding Markup

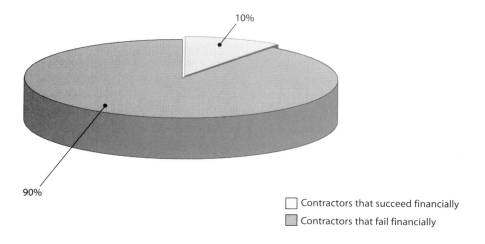

Figure 2-1
Success in construction on a 10-year cycle

do contractors continue to fail? The evidence is there, the problem areas are well defined. Why isn't this information used? In reality, there are probably as many reasons as there are people in business. But let's start with one of the most common ones.

Failing to Recognize That a Problem Exists

It's rarely the employees of a company that cause the company to fail. *Almost all failures can be traced specifically and directly to the owners of the company.* They simply don't take the time or initiative to analyze their approach to their own business. Even if they do, they don't follow up to correct the things that they're doing wrong. They continue creating the same problems, and eventually their businesses fail. In many cases, these people fall into the category of the unconscious incompetent. They don't know, and they don't know that they don't know!

The first and probably foremost problem with most business owners is their very nature. They're independent and they don't want anyone telling them what to do or how to do it. They prefer to "Do things their own way, thank you very much!" In short, *their egos get in the way of their common sense*! That cripples their decision-making process and starts the decline of the business. Even though self-determination is important to you, don't let your ego get in the way of your business. If you need some outside help for your company, get it now.

Falling Behind the Industry

The second symptom of impending business failure is the owner's failure to continue the education process that helped them to build their businesses. It's awfully easy to get comfortable and just sit back and rest on your laurels. In fact, some will argue that this is the major reason for the failure of most businesses.

Let me share a classic example of what I've found in my dealings with contractors. I once joined forces with a company called FTS Training, Inc. to set up a one-day seminar on business management for residential electrical contractors. We sent out 3,500 flyers to electrical contractors in the Portland, Oregon area. Based on a study of the unusually large number of failures among electrical contractors in the area, we thought there was a definite need for a seminar of this type.

"It's awfully easy to get comfortable and just sit back and rest on your laurels. In fact, some will argue that this is the major reason for the failure of most businesses."

We had the flyer professionally done, selected a good seminar location with easy access, and followed all the rules about mailing and proper lead-time. One of our instructors was a nationally-known author and teacher in the electrical industry. We thought he'd be a great draw for our project. We tested our presentation by giving a "tune-up" seminar to eight electrical contractors that we knew. These contractors were all successful business people, with a minimum of ten years in business. All had at least a four-year college degree, and had continued their education on a daily basis. Without exception, all eight said it was one the best seminars on business management that they had ever attended. We were ready to rescue the electrical industry! So, we sent out all those flyers. How many responses do you think we got? Two hundred? Fifty? Twenty-one? We got zero! There wasn't even one phone call asking for more details. What a bummer!

After much frustration and with mounting curiosity, we called approximately 50 of the contractors on our mailing list to find out why they hadn't responded to our invitation. We recorded what we felt at the time were absolutely amazing responses. The large majority of the electrical contractors said, "I didn't see anything in the seminar for my guys (meaning their electricians), and that's where I'm having all my problems. My crews aren't working fast enough."

What we learned was this: One, the owners didn't perceive themselves or their methods of running their companies as a problem. They thought their crews were the main reason they weren't making money. And two, the owners were almost all totally unable to pinpoint problem areas within their own companies. So we asked them direct questions about their companies, and linked the question with examples of common problems found in other companies. Most acknowledged that they had similar problems. Some also admitted, after a bit of fairly pointed questioning, that those problems were directly linked to their own actions.

"Continuing education doesn't just mean attending seminars or taking high school or college classes. What's most important is keeping up with changes in the industry."

I know of at least four companies across the U.S. and Canada that have tried to establish educational seminars specifically for the construction industry since 1993. They have had, almost without exception, about the same results that we had. Thousands of hours (not to mention dollars) have been spent developing very good educational programs. But the contractors won't spend the time or money to attend these seminars. This is one case where "if you build it, they will come" doesn't work. This anti-education syndrome, or whatever you might want to call it, is like a plague that has permeated all levels of trades in the construction industry.

Continuing education doesn't just mean attending seminars or taking high school or college classes. What's most important is keeping up with changes in the industry. That includes the day-to-day reading of technical and trade literature, especially magazines and journals that have new ideas and techniques you can use in your business. You can also learn a lot from videotapes and audiotapes that present new processes or ideas. There's a variety of educational materials available to help owners become better managers and to update business and construction methods. And they'll cost you very little in either time or money.

For example, I subscribe to a number of magazines related to both of the industries that I'm involved in, construction and computers. If you look in the cab of my pickup, you'll always find at least three audiotape programs (usually containing six or more tapes each) on business management, sales or some other topic that I listen to while driving between jobs or appointments. Why not spend your travel time learning something new about sales or business management or getting a little motivational shot in the arm rather than just passing time?

Losing Touch with the Business

A third problem is that business owners begin to get lazy and lose focus on where their company is going. They no longer work as intently, or spend their energy getting the things done that will keep their business headed in the right direction. They lose contact with their own business. Contracting is a hands-on kind of business. The excitement begins to fade when you stop taking the time to talk with employees, or go out to the job site to inspect the jobs and interact with the homeowners.

What's the result of this lack of interest in the daily job routines? The amount of labor needed to complete tasks on the various jobs tends to rise. Without the owner's involvement, the individuals working on the jobs lose their sense of urgency to get jobs done in the allotted time and within budget. The quality of work goes down, causing an increase in customer complaints and callbacks, which in turn means additional labor and material expenses. The contractor may also find that materials are mysteriously disappearing from the job sites, resulting in higher material costs. Eventually, an absentee contractor will have to deal with the biggest headache of all — irate homeowners. Usually the problems that the contractor must step in and deal with could have been prevented had he or she kept the lines of communication open and made sure the homeowners were pleased with the work as it progressed.

So what are these contractors doing if they aren't staying involved with their jobs? Apparently, they become distracted by their own perceived success! You can check this for yourself in almost any town or city in the U.S. Find a local auctioneer who specializes in selling the assets of bankrupt companies. Go to an auction on the preview day. Almost without exception, you'll find two things being auctioned off. First, you'll almost always find a boat or some other form of watercraft. Second, you'll find some kind of sports car, antique car or luxury car.

I kept track of these items at the last nine construction company bankruptcy auctions I attended. At each auction there was a boat and an expensive car. Some auctions had two or three boats and as many cars. Of course you may find some other recreational toy besides a boat. It might be a snowmobile or an ATV, but I guarantee that there'll be an expensive toy in the bankruptcy inventory. You can count on it.

Now, let's not jump to the conclusion that I think there's something wrong with playtime. On the contrary, I thoroughly believe in playing. What's wrong is that these items are included as part of the *company's* assets and not the owner's personal property. These owners tried to mix fun stuff with business stuff, and possibly beat the IRS out of a few tax dollars in the process. They misused their company funds.

Business is business, and playtime is personal. Don't mix the two. Keep your personal life and your toys out of the business. If you get sidetracked into thinking that you can play with your assets instead of working with them, you won't have them long.

Three Major Causes of Business Failure

We've just looked at some of the symptoms that show up regularly in failing businesses. But what causes the disease that these symptoms reveal — the disease of failure? There are three major reasons that contracting businesses founder. They occur in conjunction with these symptoms, but they're often harder to spot.

Failing to Understand and Apply Markup

The first reason contractors fail is that they don't understand or apply the principles governing markup. So they don't charge enough overhead and profit for the work that they do. Let me ask you the following question: How many contractors do you know that closed their business down while making a net profit of 8 percent or more? My guess is none — except for those lucky ones who retire.

Lack of Projections for Long-Term Funding

What's the second reason? Many contractors don't have the business background to be able to project cash flow and estimate the working capital they'll need to keep their business going. Every business needs adequate cash flow and an Operating Capital Reserve Account to sustain the business on a long-term basis. In my research, I found that very few contractors will take the following three necessary steps:

1. Do cash flow projections.
2. Develop a plan to obtain working capital.
3. Actually go out and put that plan to work.

I estimate that only 4 to 6 percent of all contractors in this business make written cash flow projections, or figure out how much working capital they'll need. Unfortunately, even a smaller percentage actually put out any effort to make those cash flow and working capital projections happen.

Do you know how much money you'll need today, to the penny, to run your business? How about nine weeks from today? Can you tell me what expenses you'll have then? If you don't know, do you have adequate cash reserves in the bank to pay your company expenses if your cash flow dries up? If you can't answer all these questions with an honest "yes," you've got some work to do.

Lack of Profitable Sales

The third reason for business failure is lack of profitable sales. I didn't say lack of sales, I said lack of *profitable* sales.

One of the worst traps that you can fall into in this business is *bidding jobs to be competitive*. We will not do that in our company, nor will most other successful contractors I know. I don't think anyone in our business should, and I let my prospective customers know it. Whether it comes up in a telephone conversation or on my first visit to their home or building, I tell them nicely (but firmly) that we don't do competitive bids. Saying "I'll give you a bid on a job" is dumb! It makes them think you're going to quote the job in the hope that they'll select you because your bid is the lowest. That's double dumb!

One of the mind-opening sayings I use with customers is: *"The only person dumber than an owner who goes looking for a low bid is the contractor who'll give it!"* I say this in a nice manner, not to be offensive, but rather to get the customer's attention. You have to establish your position on this subject and take control. The customer's reaction will reveal their criteria for picking a company to do their work. If a low bid is all that they're interested in, you may not want to get involved with them. We'll discuss this in greater detail later on in the book. If good quality work from a respected company is what they want, let them know that you're prepared to provide that at a fair price. But let me make my opinion on this subject very clear. I think it's better to sit home and do nothing than to take a job at a loss. If you give a "firm price quotation" to prospective customers or to general contractors for subcontract work, make sure it's at a rate high enough to guarantee a profit.

One thing that I'd like you to start doing immediately is to take a hard look at every sale your company makes. Do you have all your potential sales checked for profitability by a second person before presenting the contract to the customer? If not, why not? This should be an absolute golden rule in your company. If a job is "sold short," there's no profit to be made.

"One of the worst traps that you can fall into in this business is bidding jobs to be competitive. We will not do that in our company, nor will most other successful contractors."

Notice that each of the three reasons I consider to be a central cause for our business failures is closely related to the others. They're separate items, however, and it takes a concerted effort to remedy or prevent each one. There's no one-fix cure for them. They must be dealt with independently. There's one thing that you can do that applies to each of them, however. That's to be sure that you charge enough for the work you do. That's an excellent start.

You'll find that from time to time I'll throw in a little tidbit of information for you to think about. Here's one for you to mull over. If you're having financial problems of any kind in your business, raise the sales price of your work by 10 percent across the board, *starting today*. This applies to any area of construction that you're working in, whether it's building new homes, remodeling, or doing specialty or commercial work. It doesn't make any difference. You'll probably be raising the price of your work by at least 10 percent after reading this book anyway, so why not start enjoying that increased profit now? Besides, why wait to reduce those financial pressures?

If you believe in what I'm telling you, and you start to apply these principles, you're going to make a lot of changes in your business. One of those changes is that you'll start making some money.

Let me assure all you doubting Thomases that you can raise your prices 10 percent and it won't make one bit of difference in your percentage of sales, not one percentage point difference. If you're selling one job for every four leads now, you can raise your price 10 percent and you'll still sell one job for every four leads. Until you try it, you'll never know if it'll work for you. But by putting this one suggestion into action, I'll bet you easily profit 10 times the cost of this book in the next two weeks. Give it a try!

Now, let me ask you a question. How's your checklist for making money coming along? If you're independently wealthy, you probably have it memorized, you're living it, breathing it, and making it a part of your everyday life. If you're not independently wealthy, are you working on it? Or better, have you completed it? If you haven't even started it yet, why not? You can read this and all the other books on business management, sales, marketing, and how to build your company. But if you don't take some action and put the ideas into use, why bother to read the books at all?

Pardon me for being so direct, but it's time for another gut check. The purpose of this book is to help *you* make money. The checklist for making money is an integral part of that process. You can rationalize all you want, but again, until you've tried it, you'll never know if it works. The question then is, can you afford not to try it?

The Terminology of the Industry

Before we start a discussion of the mathematics of your business, we need to first define some of the terms that we'll use. If you find some definitions are different from those you use, try to find the common link. Don't get hung up on terms, or whose usage is right or wrong. Stay with me and look at the problem from another point of view. Find the approach that will make this work for you. That's the approach that a professional businessperson should and will take.

Job Costs

Job costs are all direct job-related expenses. If a particular expense is due to a given job, not a group of jobs, then it's a job cost. An example would be nine sheets of ACX plywood, or a white, dual-glazed, vinyl framed, clear glass, LEA, sliding glass door. These are expenses that belong to a particular job.

Overhead

Overhead is all indirect job-related expenses. Put another way, it's any expense, fixed or variable, that you can't charge directly to a particular job, but will spread out over all the jobs you do. That would include office rent or mortgage payments, office staff, a computer or related equipment, office phones, insurance, payroll taxes, a bookkeeper, and so on.

Of course you'll have to separate fixed and variable overhead, but that's not important to this discussion. To calculate the right selling price for your work, you need to use your total overhead cost, not any particular piece of it.

All of the costs your company incurs are either a job cost or an overhead cost. Besides assigning job costs to the individual job where they occur, you'll also assign a percentage of the overall overhead for the year, in proportion to your total sales for the year. If one job equals 10 percent of your total sales for the year, you'll charge that job 10 percent of the total overhead expense for the year.

Profit

Simply put, profit is what's left after all the bills have been paid. If you add up your job costs and overhead, and then subtract those two amounts from the sales price of a given job, the result is your profit.

K.I.S.

This isn't really a term, but an acronym for Keep It Simple. Many contractors are good craftspeople, but very poor business people. They tend not to want to spend a great deal of time doing paperwork and bookkeeping. So, if there's one cardinal rule in construction companies, no matter what the area, it's K.I.S. If you add the second S (for Stupid), you're getting too complicated!

A case in point: there's a lot of unnecessary confusion about the difference between job costs and overhead. Vehicles and the superintendent's time are especially disputed. Here's an easy way to look at those two items. Because it's almost impossible to accurately track how much time a company-owned vehicle is used on a particular job, or to get a superintendent to keep an accurate account of the time spent on each job, I almost always recommend that these two costs be assigned as overhead. That "keeps it simple."

Determining Your Financial Requirements

How do you determine what to charge for job costs, overhead and profit in terms of dollars and cents? Ah, that's the meat of our subject. To determine actual job costs, begin with estimating the total costs for a given job. You should be able to estimate your actual costs on any particular given job to within plus or minus 3 percent. Job costs aren't usually a major problem area. I've reviewed many new construction cost estimates that have been within ½ of 1 percent of the actual job cost. My company recently completed a dormer addition that sold for over $36,600. Our actual costs were within $30 of estimated costs. So you see, it can be done.

Remodeling work is generally more difficult to estimate, and almost always has a higher error factor than new construction. Remodeling estimates can have an error factor from as low as 1 to 3 percent, to a far more likely 10 to 20 percent. Since we began using computers in estimating, I've noticed that the error factor in our estimates is slowly coming down. I highly recommend using a computerized estimating program. It'll save you a lot of time. And you must make the best possible use of your time. That's one of the secrets of the successful folk of this world. We'll discuss computer systems in more detail later on in the book.

For the purposes of this book, let's say that estimating the job costs on an average remodeling job has an error factor of approximately 8 percent. For the average new home, it's 3 percent. These are just arbitrary numbers, but they're far closer to reality than most contractors would care to admit. There have been several studies done on the error factors in construction estimating over the last 20 years, and none has been able to pinpoint just where the major problems lie. But I've come to the conclusion, based on both my own observations and discussions with many other contractors, that the biggest problem in estimating accuracy is the *method* used to compile the estimate.

Estimating Methods

Let me share an example to illustrate this point. In 1988, I was the moderator for a class on estimating remodeling held in a large eastern city. There were about 180 contractors and tradespeople in that class, an unusually large group. As we reviewed an estimate for a room addition that we were working on in class, I realized that by the time we reached the sixth item on the estimate (of the 52 listed), there were already 180 different totals for this estimate. Granted, some of the estimates were close, but they were still all different. Of the many different estimating methods being used in the class, about 1 in 10 students was using one version or another of the W.A.G. method.

I averaged about 58 contractors in my estimating seminars. At one point I decided to keep track of the different methods of estimating that people were using in my seminars. I found that if I gave 58 contractors an identical set of plans to estimate, less than 5 percent of them would do the estimate the same way. More often, there would be 50 different approaches or styles of estimating used.

What are these different estimating methods? There are four basic methods used for estimating. Every contractor seems to have adapted his or her own variation of one of them. The basic methods are:

1. W.A.G. or wild a _ _ guess method.

2. S.W.A.G. or scientific wild a _ _ guess. This common variation of the W.A.G. involves the use of a calculator, and in some rare cases, an estimating book as well.

3. Stick Estimating is counting all the individual items and hours of labor in a given job, applying a price to each, then adding some predetermined percentage figure to cover the overhead and profit expense.

4. Unit Costing combines predetermined units of labor and materials to make up a construction component or assembly, to which you add a predetermined markup to arrive at the selling price.

Of the four methods, the unit cost method of estimating is faster and more accurate than the stick method of estimating, and considerably more accurate than the other two methods. I've shown estimators in my seminars how they can learn the unit cost estimating method in 1 to 1½ hours a day over a period of two weeks. Then they can reduce their estimating time by at least 30 to 40 percent. A good unit cost construction estimator can do a $100,000 estimate in about 2 hours, using estimating book(s), paper, pencil, and calculator. They can produce the same estimate using a good computerized estimating system in about 60 percent of the time, or 1¼ hours.

Figure 2-2 is a simple unit cost estimating form that I've filled out as an example. This form is used by a number of companies that I'm familiar with — in both new home construction and remodeling. The value of the form is in its layout. It reminds you of the steps needed to complete your estimate. To use it, you simply calculate the total amount of Labor, Materials, Subcontractor, and Other expenses that you have for each item listed and enter it in the appropriate box on the sheet. If there isn't an expense for an item or for one of the columns following the item, put a zero in the column and cross through the item so you know you haven't forgotten to check it.

You might want to put an asterisk (*) in front of the Labor column on any item that's an *allowance* amount. That way, when you transfer the information from your estimate sheet to your contract you're reminded of the allowance for the item and you can put in the *allowance limit*. You might use an allowance limit for items that the customer hasn't made a final decision on or for things like permits that have variables that will greatly affect the cost. For such items, an allowance limit reduces your liability.

Chapter 2: Understanding Markup

Job Estimate

Stone Construction Services
111 Ocean Avenue
Portland, Oregon 99999
(900) 555-1111

Estimated by: J. Peet   **Date:** 7/29/98
Customer: Litwak - Johnson Bath Remodel
Address of Job: 1233 Apple Street
Anytown, OR 99069

#	Item	Labor	Materials	Subcontractor	Other
1	General conditions plans = $125, permit = $175	40	-	300	-
2	Demolition / tearout exist. bath to studs 6/20	120	15	-	-
3	~~Excavation~~ N/A	-	-	-	-
4	~~Concrete~~ N/A	-	-	-	-
5	~~Masonry~~ N/A	-	-	-	-
6	Framing 48 S.F. - underlay = ⅝" sub = ½"	240	90	-	-
7	~~Roofing~~ 2x4 = 6/8' ply - 2/⅝" 2/1½" 8/30	-	-	-	-
8	~~Siding~~ N/A	-	-	-	-
9	Windows - replace 3°x3° w/vinyl (XO)	45	115	-	-
10	Doors trim only	-	-	-	-
11	Sheet metal vent new exhaust fan	30	10	-	-
12	Plumbing tub/lav/wc & valves (allowance)*	-	-	2150	-
13	Electrical 1 - lite/fan 3 - ⓘ 2 - ⊕ 3 - $ *	-	-	550	-
14	H.V.A.C. new floor register - white	15	15	-	-
15	Insulation / ~~weatherstripping~~ 64 S.F. ext. wall			65	
16	Drywall / ~~plaster~~ 224 S.F. walls - 48 S.F. clg.	-	-	476	-
17	~~Ceiling tile~~ N/A @1.75 S.F.	-	-	-	-
18	Cabinets 5' base 3' upper 2' linen *	-	-	1625	-
19	Surfacing 5' deck - marble @ $110	-	-	550	-
20	Tile tub/shower enclosure - 55 S.F. @ $15 *	-	-	825	-
21	Floor covering vinyl = 9 yds @ $30, 30' base *	-	-	300	-
22	Kitchen & bath accessories shwr.door,towel bar	60	40	235	-
23	~~Awning & patio~~ N/A t/p hldr, rings	-	-	-	-
24	Finish carpentry door - window trim	90	45	-	-
25	Hardware & metalwork misc. hardware	-	30	-	-
26	~~Paneling & fence~~ N/A	-	-	-	-
27	Light fixtures by owner, install 2	40	-	-	-
28	Paint & décor by owner	-	-	-	-
29	Debris removal labor, dump fee	75	-	-	35
30	Miscellaneous @ 3% job costs	23	11	212	2
	Estimated totals	778	371	7288	37

                         Job total   8474.00
                       O & P   @.46   3898.00
                         Subtotal   12372.00
                 State sales tax   @7.6%   940.28
                           Quote   13,312.28

Figure 2-2
Unit cost estimating form

Figure 2-3
Time it takes for each method of estimating

Method of Estimating and Time Involved					
Volume Estimated Job Costs	W.A.G. Minutes	S.W.A.G Minutes	Stick Method Minutes	Unit Cost Minutes	Unit Cost with Computer Minutes
0 to $10,000	1+	3 to 5	55 to 70	45 to 60	25 to 35
$10 to $25,000	1	5 to 7	80 to 95	60 to 75	35 to 45
$25 to $50,000	2	7 to 9	100 to 125	75 to 85	45 to 55
$50 to $75,000	3	9 to 10	130 to 145	85 to 100	55 to 65
$75 to $100,000	4	10 to 11	150 to 165	100 to 115	65 to 75
$100 to $125,000	5	11 to 12	165 to 185	115 to 125	75 to 85
$125 to $150,000	5	12 to 13	185 to 200	125 to 140	85 to 95

When you've filled down all the items, total up the columns and combine those totals to come up with your Job Total. To this you add your Overhead and Profit (markup) to come up with a subtotal to which you can add your state sales tax if applicable. The final total is your estimate or quote for the job.

I frequently use this form to organize my thoughts and to take notes on when reviewing a job with the customer. When I return to my office, I enter the information into my computer estimating program and have the estimate done in a fraction of the time it would take me to do it by hand. I've included a blank copy of the form in the back of the book. Make copies to use as it is, or change it to suit your business. Remember to always have someone else in your office go over your estimate before you give the quote to the customer.

Figure 2-3 shows the approximate time it takes to estimate job costs using each of the four methods listed. Of course the times will vary widely, depending on the individual, their experience, the reference materials used, and their skill with a calculator and/or computer.

For Those Stuck on Stick Estimating

One of the reasons estimators prefer stick estimating is that when their estimate is finished, they have a complete material list ready for the job. They say, "I have all my materials listed. If I get the job, we're ready to go!" For those who are wedded to stick estimating, here's a point to consider. Why is a material list important, or even necessary, if you haven't got a contract yet? Why waste the time making a list unless you know you have the job?

Here's a little sales advice. An estimate is only part of the process of getting work for your company. The estimate is important, no doubt. We can all agree on that. When you make a presentation or proposal to a potential customer, they'll either agree to buy from you or not, for whatever reason. If they do decide to buy from you, what they're actually buying (in order of importance) is:

1. you
2. your company
3. the job and all its component parts

They don't care about your material list. Several studies done by the NAHB, other trade associations, and a number of consumer groups have shown that price ranks sixth or seventh on the list of things that determine from whom a customer will buy. Focus your time on selling yourself and your company. If you know your business, and you dress and act professionally, your customer will pick up all the other details that are important to them. Arrive at a good design, then a firm price, then spend the time that you used to waste on stick estimating to sell that job.

If most estimator/salespersons would spend the time they use doing material takeoffs on improving their selling techniques instead, they could increase their sales by 5 to 10 percent a year. And if they switched to unit costing, I think they'd probably increase their estimating accuracy by about the same percentage. This is just another tidbit for you to consider.

Figuring Overhead Expenses

We've defined job costs, overhead, profit, and I've recommended a method of estimating. Next you need to determine what the financial requirements are for your company. Then when you build a job for a customer, you'll know that you'll have enough money to cover the job costs, overhead expenses, and most importantly, *your profit!*

What! Did somebody just say, "I don't have any overhead, I work out of my home"? Don't you believe it! I know that there are still people out there who actually believe that running a business from an office in their home somehow magically reduces their overhead expenses to zero. They believe that those expenses can be absorbed in the family budget. If you have such a thought in your head, please go back and review the definition of overhead.

Everyone in the construction business who owns their business has overhead expenses. If you're in business, you have overhead. I have included some charts that I've compiled over the years that list the minimum overhead expense items for companies involved in the construction business. These list the *minimum expenses* that we all have in this business, regardless of whether we build new buildings, remodel existing buildings, or we're specialty contractors like electricians, plumbers, drywall installers, roofers, or floor covering installers.

I've used these charts in several workbooks for seminars that I've given in the U.S. and Canada. Based on continuing feedback, the numbers are fairly accurate, and they've held up to the test of time. As I develop the theory of markup, we'll refer back to the charts from time to time.

These numbers will hold true for both new construction and remodeling, regardless of your location. If you've been in business for a while, the numbers for your company may be a little different, but the percentages should be about the same. If you want to fill in the blanks so you focus on what we're doing, then use the numbers that suit your company. Look these charts over carefully, and keep the following in mind:

- The list of overhead expenses shown here, while incomplete, should cover the majority of your expenses.

- The highs and lows may not exactly match those of your company. These should be viewed as starting points, providing you with an approximation of your expenses. You must determine your exact expenses and their percentage of your total sales.

- You may decide as you look over the items that you want to combine several of them to give a more accurate picture of your company expenses. A good example of this might be that you own the company but you also do sales for the company. As the charts show, those are two different expenses. You may decide that instead of paying yourself an Owner's salary at 5 percent and Sales commissions at 6 percent, that you combine the two under Owner's salary at 10 percent. That immediately reduces both your overhead expenses and your markup. If you also do Job supervision (minimum expense of 4 percent), then you might add that in and pay yourself 12 percent. However, if you also do some of the carpentry on the job, then you should pay yourself a salary at an hourly rate just as you would pay any other carpenter that you had to hire. That amount should be included in your job costs for the particular job and figured into your estimate. It's not figured as part of your overhead on this chart. So now we have you with a salary package at 12 percent on the chart. Why not the whole 15 percent as the chart would indicate? Because you're not doing each of these jobs full time, nor would it take a full-time commitment to do these jobs in a company the size we're dealing with here. So, you can reduce your salary by a small percentage and still be compensated at a fair rate for your work. You're still getting paid for everything you do, as I think you should, but not so much that it would increase your prices and remove your company from consideration by your customers. This is a tough issue to deal with. It has nothing to do with "bidding" or "what the market will bear" but rather with charging a "fair" (neither too little nor too much) amount for the work you do as both an owner and an employee of your company.

- After reviewing the charts in Figures 2-4, 2-5 and 2-6, you can use the blank Overhead Expense Chart included in the back of the book to list your overhead expenses. Make copies so you can update your charts on a regular basis.

There's only one line included on the charts for taxes. You'll need to insert the taxes that are appropriate for the state and locale where you do business. Of course, payroll taxes should be included as part of the overhead Salary expense (Workers' Compensation, Social Security, Medicare, etc.), but only for employees who are part of your company staff and are overhead employees (owner, bookkeeper, sales staff, etc.). Field employee payroll taxes are part of your job costs and should be included in the labor/salary expense for each job. Sales taxes are a separate issue and should be handled as required by the state in which you do business.

You'll note that the overhead in the charts ranges from 15.45 to 53.0 percent. That range should convince you that no single markup figure could possibly work for all companies. And notice that these figures are in direct contrast to the myth that a contractor can operate on 10 percent overhead and 10 percent profit. I'm not going to tell you that the math, correctly done for your company, may not dictate a markup of 1.21, which would justify the 10 percent and 10 percent figures. (121 is the figure you arrive at if you take a number, say 100, and multiply it by 110 percent and then 110 percent again: $100 \times 1.10 = 110$; $110 \times 1.10 = 121$, which gives you a 1.21 markup.) What I am saying is that it's a rare case when the numbers do come out exactly 10 percent and 10 percent. So when you hear the statement made by so many people that contractors should "all" use a blanket 10 percent each to calculate their overhead and profit margins, you'll know that this simply is not true.

This 10 percent overhead and profit myth is perpetuated by a lot of "knowledgeable people" who work in industries associated with or on the fringes of the construction industry. You'll find them in government agencies, insurance companies, banks, law firms and most often in the media. The problem is that these folks fail to take into consideration that all contractors don't do the same volume of business each year, nor do they have the same overhead expense. So your numbers are undoubtedly different from anyone else you know. And your markup should be different, also.

Few contractors ever take the time to really consider that last statement. They listen to what the experts say they should be charging for their work. If "industry experts" tried to tell me what I should be using as a markup, I'd ask them, "How could you arrive at that number, not knowing my company overhead or what percentage of profit I want to make?" I know they

Overhead Expense Chart
General Contractor — Remodeling
(For every $100,000 in annual volume sold, built and collected)

Overhead Item	Low percent	High percent	Low expense	High expense
1. Advertising	1.50	5.00	$1,500	$5,000
2. Sales	5.00	8.00	$5,000	$8,000
3. Office Expenses				
Staff	3.00	7.00	$3,000	$7,000
Rent	0.35	1.20	$350	$1,200
Office equipment	0.10	0.50	$100	$500
Telephone	0.20	0.75	$200	$750
Computer*	0.10	0.35	$100	$350
Office supplies	0.05	0.20	$50	$200
4. Job Expenses				
Vehicles	0.75	3.00	$750	$3,000
Job supervision	4.00	6.00	$4,000	$6,000
Tools & equipment	0.20	0.75	$200	$750
Service & callbacks	0.10	0.50	$100	$500
Mobile telephone	0.05	0.30	$50	$300
Pagers	0.05	0.15	$50	$150
5. General Expenses				
Owner's salary	6.00	8.00	$6,000	$8,000
General insurance	0.25	1.50	$250	$1,500
O.C.R.A.**	1.00	4.00	$1,000	$4,000
Interest	0.05	0.75	$50	$750
Taxes***	0.00	3.00	0	$3,000
Bad debts	0.00	0.30	0	$300
Licenses & fees	0.10	0.25	$100	$250
Accounting fees	0.15	0.30	$150	$300
Legal fees	0.15	0.50	$150	$500
Education & training	0.15	0.30	$150	$300
Entertainment	0.10	0.20	$100	$200
Association fees	0.10	0.20	$100	$200
TOTALS	**23.50%**	**53.00%**	**$23,500**	**$53,000**

* Computer expense = hardware, software and support services.
** Operating Capital Reserve Account. You may reduce this by 70% to 90% to maintain the account once you've reached the goal for the account.
*** State and local taxes will vary widely with your location and business structure. Federal taxes are not included.

Figure 2-4
Overhead percentages for general contractors in remodeling

Chapter 2: Understanding Markup

Overhead Expense Chart
General Contractor — New Home Construction
(For every $100,000 in annual volume sold, built and collected)

Overhead Item	Low percent	High percent	Low expense	High expense
1. Advertising	1.00	2.00	$1,000	$2,000
2. Sales	2.00	5.00	$2,000	$5,000
3. Office Expenses				
Staff	1.50	3.50	$1,500	$3,500
Rent	0.35	1.20	$350	$1,200
Office equipment	0.10	0.50	$100	$500
Telephone	0.10	0.75	$100	$750
Computer*	0.10	0.35	$100	$350
Office supplies	0.05	0.20	$50	$200
4. Job Expenses				
Vehicles	0.75	2.50	$750	$2,500
Job supervision	1.50	2.50	$1,500	$2,500
Tools & equipment	0.15	0.50	$150	$500
Service & callbacks	0.20	0.40	$200	$400
Mobile telephone	0.05	0.30	$50	$300
Pagers	0.05	0.15	$50	$150
5. General Expenses				
Owner's salary	4.00	6.00	$4,000	$6,000
General insurance	0.25	1.00	$250	$1,000
O.C.R.A.**	1.00	4.00	$1,000	$4,000
Interest	1.75	3.00	$1,750	$3,000
Taxes***	0.00	3.00	0	$3,000
Bad debts	0.00	0.15	0	$150
Licenses & fees	0.10	0.25	$100	$250
Accounting fees	0.10	0.25	$100	$250
Legal fees	0.15	0.30	$150	$300
Education & training	0.10	0.20	$100	$200
Entertainment	0.05	0.10	$50	$100
Association fees	0.05	0.10	$50	$100
TOTALS	15.45%	38.20%	$15,450	$38,200

* Computer expense = hardware, software and support services.
** Operating Capital Reserve Account. You may reduce this by 70% to 90% to maintain the account once you've reached the goal for the account.
*** State and local taxes will vary widely with your location and business structure. Federal taxes are not included.

Figure 2-5
Overhead percentages for general contractors in new construction

Overhead Expense Chart
Specialty Contractor
(For every $100,000 in annual volume sold, built and collected)

Overhead Item	Low percent	High percent	Low expense	High expense
1. Advertising	1.00	2.00	$1,000	$2,000
2. Sales	1.00	2.00	$1,000	$2,000
3. Office Expenses				
Staff	3.00	7.00	$3,000	$7,000
Rent	0.35	1.20	$350	$1,200
Office equipment	0.10	0.50	$100	$500
Telephone	0.25	0.75	$250	$750
Computer*	0.10	0.35	$100	$350
Office supplies	0.05	0.20	$50	$200
4. Job Expenses				
Vehicles	0.75	3.00	$750	$3,000
Job supervision	3.00	5.00	$3,000	$5,000
Tools & equipment	0.20	0.75	$200	$750
Service & callbacks	0.10	0.50	$100	$500
Mobile telephone	0.05	0.30	$50	$300
Pagers	0.05	0.15	$50	$150
5. General Expenses				
Owner's salary	6.00	8.00	$6,000	$8,000
General insurance	0.25	1.50	$250	$1,500
O.C.R.A.**	1.00	4.00	$1,000	$4,000
Interest	0.50	0.75	$500	$750
Taxes***	0.00	3.00	0	$3,000
Bad debts	0.00	0.30	0	$300
Licenses & fees	0.10	0.25	$100	$250
Accounting fees	0.15	0.30	$150	$300
Legal fees	0.15	0.40	$150	$400
Education & training	0.15	0.30	$150	$300
Entertainment	0.10	0.20	$100	$200
Association fees	0.10	0.20	$100	$200
TOTALS	**18.50%**	**42.90%**	**$18,500**	**$42,900**

* Computer expense = hardware, software and support services.
** Operating Capital Reserve Account. You may reduce this by 70% to 90% to maintain the account once you've reached the goal for the account.
*** State and local taxes will vary widely with your location and business structure. Federal taxes are not included.

Figure 2-6
Overhead percentages for specialty contractors

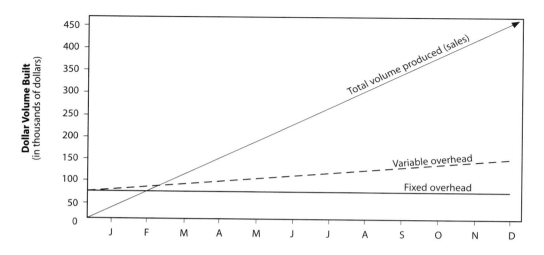

Figure 2-7
Comparison of overhead to dollar volume built

couldn't tell me. That's well-meaning mischief at its worst. With all these "industry experts" trying to tell us how to run our businesses, is it any wonder that the failure rate among contractors is so high?

The graph in Figure 2-7 shows the comparison between fixed and variable overhead expenses and dollar volume. While it may seem simple, it's pictorial proof that as your dollar volume of sales goes up, your overhead also goes up, not down as so many industry experts seem to believe.

The straight line across the bottom portion of the graph is labeled *Fixed overhead* and represents your routine, fixed expenses for your business. These are your rent, insurance payments, association dues, licenses, office staff, etc. These items remain pretty close to the same throughout the year. The dotted line represents your *Variable overhead* expenses, which include your Operating Capital Reserve Account, job supervision, tools and equipment, sales commissions, etc. You may have a difference of opinion about where some of these items belong, but that isn't important right now. What we're looking at is the fact that as your *Total volume produced* (your sales) goes up, your variable overhead also goes up. The two are tied together. So even though your fixed items of overhead may remain constant, the variable items go up with your volume of sales, thus raising your total overhead expense. Let's clarify this point a little further by looking at some examples.

Suppose you predict that your fixed overhead expense is going to be $50,000 and your variable overhead expense is going to be 30 percent of your sales for a given year.

- With sales of $75,000, your overhead is $50,000 + $22,500 (30% of $75,000 = $22,500) for a total of $72,500.
   $72,500 is 96.67% of $75,000.

- With sales of $150,000, your overhead is $50,000 + $45,000 (30% of $150,000 = $45,000) for a total of $95,000.

    $95,000 is 63.33 % of $150,000.

- With sales of $300,000, your overhead is $50,000 + $90,000 (30% of $300,000 = $90,000) for a total of $140,000.

    $140,000 is 46.67% of $300,000.

- With sales of $450,000, your overhead is $50,000 + $135,000 (30% of $450,000 is $135,000) for a total of $185,000.

    $185,000 is 41.11% of $450,000.

You can also see that as your sales increase, your overall overhead expense becomes a smaller percentage of your total sales volume and the gap between the two grows wider on the graph. But overhead is still going up, not down! The next time you start thinking about cutting your markup because you're selling more, remember, "As volume goes up, expenses also go up."

Start-Up Overhead

Let's take a look at some of the basic expenses that you'll have when you start up a new company. Even if you've had your business for a while, you'll find this review helpful. It'll give you a background for some of the theories that we'll be covering as we go along.

Location Selection — How do you know if a given location will support your company? Will there be enough jobs in your county, city, or neighborhood to support the business? This is easy to calculate; however, finding the information necessary to make the calculations isn't always so easy.

You'll need to go to the building departments of the areas that you'll be doing business in and find out the total dollar figure for all residential construction, both new and remodeling, that was permitted in the last 12 months. Few city or county governments normally separate the total dollar volume of permits for new construction and remodeling. A conservative estimate would be that the dollar volume for residential remodeling work is about the same as the dollar volume for new home construction. Actually, there's been a little more remodeling work lately, but let's take the conservative approach and call them equal.

The permitted work total includes work that hasn't been completed yet. So divide the total in half to find the work that was actually sold and built. That will give you the approximate amount of work that was done in the area for both remodeling and new homes. Now, divide that number in half again to separate the amount for remodeling from the amount for new home construction. That's about 25 percent of the total you began with. The actual amount will probably be slightly higher due to the work that was done without permits, but it's best to be a little cautious in your calculations. It's always better to find that there's more rather than less work available than you've anticipated.

Your next step is to find the population of the area or areas that you want to work in. Divide the total construction dollars by the population to find the amount of work done per person. Then you can predict about what percentage of the total construction "pie" you can expect to sell and build. For example, in any particular metropolitan area, there will be an average of $300 to $400 of work done per person each year. The share of the pie will vary considerably between the contractors specializing in new home construction, remodeling or commercial work, or specialty jobs because of the difference in the average job size for each type of work.

Another way to determine the "pie" available is to talk with several other contractors already in business. Find out what volume of work they completed during the previous 12 months. That'll help you determine what kind of volume your area will support. You'll find that the professional contractor is more than willing to share this kind of information, so don't be afraid to ask. They know that sharing information helps make the construction business a better place for everyone. You'll occasionally run into jerks that won't tell you anything, or who might even lie to you! With a little experience you should be able to weed them out.

Projected Volume — The average company is doing 1 to 3 percent of the total area volume after their third year in business. The first year in business, most remodeling companies can expect to complete a volume of $150,000 to $300,000, more if they have a heavy emphasis on sales. Everyone measures success on a different yardstick. Some people want their company to grow after the first year, and others don't. If you decide to grow and expand your business, you'll probably grow at about 20 to 30 percent during the next two to three years. After that, a steady 6 to 10 percent growth per year is considered normal.

If you decide to keep your company small, then you may want to limit your growth to some specific percentage each year. Keep economic factors in mind as you set goals for your company. History has shown that the dollar loses a little in value each year, so you need to raise your volume by an equivalent percentage just to stay even.

A word of caution here: Be careful not to outgrow (oversell) your company's ability to perform. Keep a sharp eye on your people and their ability to get the job done on schedule. At some point, the "Peter Principle" will kick in. Based on the failure rate in this business, Peter must be setting a new record for asset kicking each year among contractors.

A remodeling contractor will normally have an average job size of $7,000 to $9,000 during the first year or two in business. That seems to be a fairly common range for new contractors around the U.S. A contractor building new homes will have an average job size of $130,000 to $160,000.

So the new home contractor could build from $260,000 to $2,000,000 or more the first year. The growth rate for new home contracting depends on the type of homes the contractor builds and the aggressiveness of the company's marketing and sales plans.

If your company will be doing commercial work, new or remodeling, the type of work you've selected to do will determine the volume. You'll have to do some research here. If you can't find anything about the volume of work you can expect in the area, set a goal of $250,000 to $300,000 in sales for the first year in business.

Specialty contractors generate a smaller volume of sales than general contractors. But a specialty contractor who's been in business for a decade will certainly have more business than the small one-man general contractor just starting out. The specialty, the company marketing program, and the number of years they've been in business will determine the dollar volume.

The Office — Once you've decided on a location, you may have to locate office space. Look for a suitable building available at a rent that's no more than 0.35 to 1.5 percent of your total projected volume for the year. If you can, however, I recommend that you start and run your business from your home, at least for the first couple of years. Your volume of business sold, built and collected usually won't support the overhead that goes with a commercial office space when you first start out.

I also suggest you wait if you're considering having a showroom. You'll have enough problems developing business at an adequate profit margin in the beginning. A showroom requires a location close to the buying public, and that can be expensive. Besides the maintenance and overhead, a showroom must be staffed, and that expense can be a real anchor if you're small and just getting started. If you're worried about the perception your potential customers will have of your company if you don't have a showroom, that's easy to handle. When a customer asks me where our showroom is, I simply tell them, "We run our business from an office in my home. It's more efficient for us and we can pass that savings on to our customers. A number of our suppliers have showrooms set up and they encourage us to bring our customers in to see the materials that we'll be using in their homes. Would you like to visit a few?" If you phrase it that way, you'll find that their concerns will go away. Most people just want to know that you're going to be around a while. If they can go to a showroom and talk to suppliers who know you and your business, and they can see for themselves the type of materials you'll be using, they'll be quite satisfied.

Some new-home builders like to locate their business offices in a model home. That's good if you can afford to keep the model off the market. Remember, you have to make payments on it while it sits. If you can swing it financially, a model home offers a good opportunity for you to show your work to your prospective customers.

Always locate your business office within easy access to the major roadways or freeways in the area where you're doing business. Travel time to and from jobs in major metropolitan areas can range from 15 minutes to two hours a day. A long commute can make a big dent in your profits if you don't account for it in the job estimate or in your overhead expense projections. There's a time-proven formula for figuring how to charge travel time to the job in Chapter 7. We'll cover the subject in more detail at that time.

Here's something else to consider when you choose a location for your office. Its location will affect the cost of your rent, utilities, outdoor advertising, fuel for your vehicles, and the cost of insurance for the building and its contents, as well as the vehicles registered at that location. It will affect almost anything you do. Check into these costs carefully before you settle on an office location. Of course, if your office is in your home, the savings in rent will help offset some of these other costs.

Ratio of Employees to Dollar Volume — Let's discuss one of the fatal traps too many contractors fall into during the first couple of years in business. It's common for contractors to employ more people than they need to complete the work they're doing. More, at least, than a young company can afford to support.

Some contractors like to have employees so all their work is done "in house." They believe they can produce a better quality job, and meet a tighter time schedule, if they have "their own people." The opposite approach, of course, is to use as many subcontractors as possible. The obvious benefit to that is shifting the financial risk to the subcontractors. If a sub takes more time or uses more materials to complete the job than estimated, that's their problem and not yours.

I've found that few contractors sit on the fence on this issue. They have very strong opinions one way or the other. Whichever you choose, there are some numbers that you must keep in mind. I've done considerable research on these numbers, but don't take my word for it. Think it through and decide what will work for your company. Don't stick to your present method just because that's the way you've always done it. *Choose the method that will give you the highest percentage of profit at the end of the year!*

You can use this list as checklist to see whether you have the correct ratio of employees to the dollar volume you build and collect.

- If you're a remodeling contractor who uses subcontractors for all or nearly all of your work, the ratio is one employee for every $350,000 to $375,000 work sold, built and collected.

- If you have two or three employees, that volume per employee may decline to $275,000 to $315,000.

- If you have four, five or six employees, your numbers may decline a bit further. Your volume per employee with six full-time people may slip down to about $185,000.

- If you employ seven, eight or more people, your ratio may decline to $160,000, or possibly even a little lower.

Look at the graph in Figure 2-8. You can see that once you have six or seven employees, the curve tends to flatten out. The volume of work that your company produces per employee shouldn't vary more than plus or minus 5 percent from the numbers on the graph.

I based the graph on information published annually in *Remodeling* magazine. Each year, the magazine puts together a list of 50 contractors who have done well (in their editor's opinion) during the past year or two. The list includes the company's dollar volume for the previous year, and the number of employees they have. Assuming that these companies are providing accurate information, I've come up with the figures for the average of dollar volume built to the number of employees for successful businesses. I always toss out the highs and lows of the group, and base the averages on about 30 of the contractors that have been included in the article. Over a period of several years, these figures seem to hold fairly steady even though the companies on the list may change.

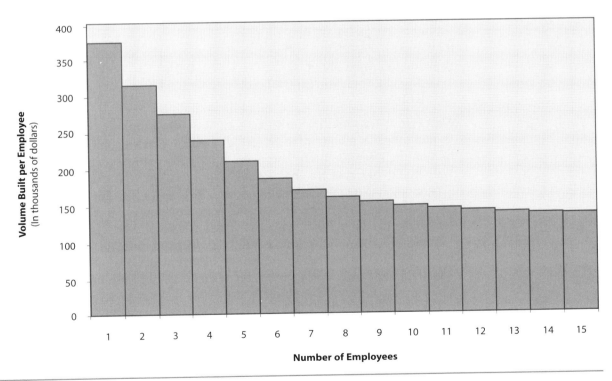

Figure 2-8
Volume built per employee for residential remodeling

I should mention that these numbers are for general remodeling contractors doing all kinds of jobs, not just specialty items like window replacement, siding or roofing. There's a similar production graph for companies doing new home building on page 58.

I think that the numbers from the magazine tend to be a little better than I'd consider normal, but that's why the magazine chose these companies. Anyway, they'll serve as a yardstick so you can compare your company's numbers. If you study the graph in Figure 2-8, you'll have the best visual yardstick available for measuring what your production should be.

I've talked with contractors who are doing a $375,000 total dollar volume per year and employ three carpenters, one helper, two superintendents and two people in the office. If you include the owner of the company, that's a total of nine employees. And, in a word, that's suicide! It's simply a matter of time before the financial crunch hits companies like these, and they disappear.

"If you've let your production per employee slip below the average (and you decide to make the changes that I recommend), you're in for a real battle with your staff."

Let's take a look at the numbers. For purposes of this illustration, we'll say that job costs are 58 percent, overhead is 34 percent and net profit is, of course, 8 percent. These are fairly typical numbers for a small remodeling company. Then $375,000 volume means that the job costs at 58 percent are estimated to be about $217,500, which leaves $157,500 to cover overhead and profit. Overhead at 34 percent is $127,500 and 8 percent profit equals $30,000.

Now, let's add up some individual overhead numbers. Let's say advertising for this company is running at 4 percent, sales commissions are 5 percent, the owner's salary is 6 percent (which is at least 2 percent lower than it should be), rent for the office is 1 percent, and the two office staff are 6 percent. (At 6 percent, that's about $940 per month for each. How many competent office employees would actually work for that pay?) Let's throw in state and local taxes and other fees of at least 3 percent, and you have a minimum overhead expense of 25 percent or $93,750. That leaves just $63,750 to cover the remainder of the overhead and all the profit. We've established that a company needs to make a minimum of at least 8 percent net profit to survive, which is $30,000, so you now have a balance of $33,750 to pay all the rest of the overhead for the year.

Go back and look at the chart for Overhead for General Contractors in Remodeling in Figure 2-4. Where does the money come from to pay two job superintendents and the balance of the overhead items listed? Even assuming there aren't any problems on the jobs during the next year, no late collections, no callbacks or service calls, and everything runs perfect, we're still short — and you know that there's no such thing as a perfect day, week, month or year in the construction business! You can see that even though the numbers shown here are rough, you can't operate with this volume, have this many employees and still make a minimum profit. A company operating like this has some very serious problems that need to be resolved. Something's got to go!

The Ratio of Employees to Dollar Volume in Established Companies

If you're doing residential remodeling or you're a specialty contractor, the following advice is for you. When your volume of jobs sold, built and collected divided by the total number of employees in your company, including yourself, drops below the comparisons shown on the chart in Figure 2-8, it's time to lean up and clean up — and fast. Even if you start right away, it will take three to six months for the financial crunch to subside. And that's assuming that your production volume remains in line with the chart.

The Natural Resistance to Change — A word of caution here for all contractors. If you've let your production per employee slip below the average (and you decide to make the changes that I recommend), you're in for a real battle with your staff. All of a sudden you're going to be telling them that their production is going to have to increase or the company can't survive.

You won't believe the excuses you'll get, the crying you'll hear, the whimpering, the wailing, and the gnashing of teeth! You'll have to duck the buckets of B.S. thrown your way. Depending on your staff's creativity, you may have *truckloads* thrown at you. Everyone will claim to be working at peak efficiency already. How can you possibly ask for more? They'll declare that it would be easier to walk on water than get more production out of your already overworked and overburdened staff!

What if you decide you have to cut back your staff? Well, that's the end of your nice-guy status! How can you do such a thing, especially when it's only seven months until Christmas? It's hard to take, but unless you increase your production, your company will die, taking everyone with it.

Increasing Production — Put a plan together that will gradually get production per employee up to where it needs to be for your company to survive. It might take as long as six months to get from where you are to where you need to be. Here's an approach you might use to make it happen.

First, hold a conference with your key people to lay out a plan of action. If you have good people, it won't take them long to see where you're headed and the necessity for change. If they want to keep their jobs, they'll get behind what needs to be done. Once you have a plan laid out, call a meeting of the whole staff, including your field people, and introduce the plan to them. You must be frank, direct, and to the point. This isn't the time for finger pointing, name calling, reprimands or threats. This type of negative behavior is counterproductive. It's time for all of your people to get going in the same direction, and all together.

Part of your plan should include a system of regular checks to ensure that the plan is working and you're on the right track. The important thing is to stick to your goal of getting the volume per employee at a profitable level. Will it stay on track throughout this process? Nope, and you already know that. Will it be easy? Nope, and you know that, too. When you find something or someone that's not working the way you want, take action. No plan is 100 percent perfect from the start. But with a few minor corrections as you go along, you'll get to your destination.

"The bottom line is this: For everyone's sake, you must get the volume built per employee back in line."

What if there are employees still dragging their feet, who aren't willing to correct or redirect their efforts toward the goal? It's time to give them a transfer to your competition. Let them help the another company go broke. The bottom line is this: For everyone's sake, you must get the volume built per employee back in line.

One final word here on the corrective process. In my travels, I've talked with many contractors who are absolutely positive that they're getting the maximum production out of each employee. They don't believe that they can increase their production. If you've fallen into that trap, let me assure you, you're perfectly normal. But I'll bet you really don't know what you and your staff are capable of producing. And you won't until you seek out information from other good companies regarding their production and compare their production levels to your company's.

You need outside input and information from other companies and/or individuals who really know what they're doing. Find someone who knows this stuff, and have him or her look at your business and show you some different approaches. You may have to pay for this advice, but it'll be worth it if they can show you what you need to do, or at least get you headed in the right direction. Put your ego in your pocket, and sit up and listen. This is one of those times when you can feed your ego or you can feed your family, but you can't feed them both.

A well-organized remodeling contractor should be able to produce *at least* $700,000 to $850,000 a year in volume sold, built and collected with one person in the office and one field person or superintendent. By well organized I mean a contractor who has effectively automated his company with a good computer system, including industry specific software, and an efficient phone/pager/messaging system. With this basic equipment and one or two steady employees, you could make even more in many cases. I know one remodeling contractor who builds well over $1,500,000 each year all by himself, with no employees — and he's been doing this for several years. I know others who build from $800,000 to $1,000,000 annually with only one or two employees. It's certainly possible.

Now, don't go jumping to conclusions. Firing all your employees and hiring a bunch of subs won't make you rich. It doesn't work that way! But you do need to take a hard look at how you conduct your business, and the best ratio of work produced per employee. Some contractors are very good at getting top performance and high production from their employees. They also seem to be able to handle the increased load of bureaucratic nonsense heaped on them by various local, state, and federal agencies that goes along with having your own employees. Others, like me, aren't good at inspiring employees and are even worse at dealing with paperwork. We use subcontractors.

Not surprisingly, some would argue about my numbers, and that's OK. Here's a way of checking to see where the correct number lies for your company. If your company meets the following criteria, then whatever number of employees you have is fine. Your ratios are working for you. If it ain't broke, don't fix it.

Checking Your Numbers — How does your company measure up in each of the following areas?

Owner's salary: The owner of the company must be paid a salary. No, I'm not talking about company profits; I'm talking take-home pay. You need to issue yourself a paycheck the same as you do for your employees — just like you used to get when you were working for that other guy! You are an employee of your company. Profit and the owner's salary are two entirely different items.

Most, and I really do mean most, contractors don't take a salary out of their company. They mistakenly believe that the company profits are their salary. Profit belongs to the company, and should be used for reinvestment, growth and new opportunities for the company. The owner's salary is the money that the company pays you to run the company effectively. Your salary, regardless of the type construction you do, should be at least 6 percent of the company's gross sales for the year, preferably 8 percent or more. Time for another gut check. Are you taking a regular salary from your company? If you are, that's terrific. If you're not, why not?

Owner paid for all work on jobs: In addition to an owner's salary, if you physically work on a job, any job and for whatever reason, pay yourself the same amount you'd pay another employee. That includes concrete work, framing, siding, hanging doors and windows, deliveries or anything else. If you're working on a job, you get paid.

Spouse paid: If your spouse or any other family member is doing bookkeeping, answering phones, or taking care of any other company-related responsibility, pay them either hourly or a salary. The old nonsense of "I can't afford to pay my wife to do our books" just doesn't wash in a professionally-run company. Would you work for someone without pay? Assuming you wouldn't, how could you ask your wife or any other family member to work for nothing? It isn't fair, or good business, to ask someone to work for your company for nothing, even if they say they don't mind.

A net profit: Is your company making a net profit of at least 8 percent?

If your company is meeting all the above criteria on a regular basis, then you've got the right number of employees working in your company. If you're not meeting the minimum profit figure of 8 percent, it's time to do something about it.

Quarterly Overhead Review Forms

The forms in Figures 2-9, 2-10 and 2-11 closely parallel the overhead expense charts that we looked at earlier in this chapter. You can use them in a variety of ways, whether you're just getting started or you've been in business a while. I've designed them for two purposes: to help you project present overhead expenses, and to keep track of past expenses. If you use them consistently, they'll help you keep track of where you are, financially. I've filled them out so that you can get an idea of how they look and what they can tell you about your company. You can see that some items run slightly higher or slightly lower than the average percentages in the Low/High Percents column. It's normal for items like legal or accounting fees, or bad debts to go over or under in any given quarter. However, averaged over time, from quarter to quarter, they should fall between the percentages shown on the charts.

If you install a copy of one of these forms in a good spreadsheet program on your computer, you can look at your company financial history and future projections in seconds. You can see what your company is doing, spot problems, and make decisions using information that's simply not available to those who haven't yet automated their companies. There are blank copies of these forms in the back of the book.

Quarterly Overhead Review
General Contractor — Remodeling
(Computations based on annual volume built and collected)

Overhead Item	Low / High Percents	Last Quarter	Percent	This Quarter	Percent
1. Advertising	1.50 to 5.00	$2,023	2.94	$2,338	2.99
2. Sales	5.00 to 8.00	$5,145	7.50	$5,846	7.50
3. Office Expenses					
Staff	3.00 to 7.00	$4,699	6.85	$4,699	6.02
Rent	0.35 to 1.20	$515	0.76	$515	0.66
Office equipment	0.10 to 0.50	$137	0.20	$47	0.06
Telephone	0.20 to 0.75	$425	0.62	$492	0.63
Computer*	0.10 to 0.35	$123	0.18	$49	0.07
Office supplies	0.05 to 0.20	$41	0.06	$113	0.14
4. Job Expenses					
Vehicles	0.75 to 3.00	$583	0.85	$619	0.79
Job supervision	4.00 to 6.00	$4,116	6.00	$4,677	6.00
Tools & equipment	0.20 to 0.75	$295	0.43	$74	0.09
Service & callbacks	0.10 to 0.50	$75	0.11	$461	0.60
Mobile telephone	0.05 to 0.30	$144	0.21	$171	0.22
Pagers	0.05 to 0.15	$62	0.09	$62	0.08
5. General Expenses					
Owner's salary	6.00 to 8.00	$4,802	7.00	$5,457	7.00
General insurance	0.25 to 1.50	$878	1.28	$1,013	1.29
O.C.R.A.**	1.00 to 4.00	$2,058	3.00	$2,339	3.00
Interest	0.05 to 0.75	0	0.00	0	0.00
Taxes***	0.00 to 3.00	$480	0.70	$561	0.71
Bad debts	0.00 to 0.30	0	0.00	0	0.00
Licenses & fees	0.10 to 0.25	$82	0.12	$75	0.10
Accounting fees	0.15 to 0.30	$123	0.18	$290	0.38
Legal fees	0.15 to 0.50	0	0.00	$150	0.20
Education & training	0.15 to 0.30	$69	0.10	$125	0.17
Entertainment	0.10 to 0.20	$110	0.16	$84	0.11
Association fees	0.10 to 0.20	0	0.00	$45	0.06
TOTALS	23.50% to 53.00%	$26,985	39.34%	$30,302	38.87%

* Computer expense = hardware, software and support services.
** Operating Capital Reserve Account. You may reduce this by 70% to 90% to maintain the account once you've reached the goal for the account.
*** State and local taxes will vary widely with your location and business structure. Federal taxes are not included.

Figure 2-9
Quarterly overhead review form for a typical remodeling company

Chapter 2: Understanding Markup

Quarterly Overhead Review
General Contractor — New Home Construction
(Computations based on annual volume built and collected)

Overhead Item	Low / High Percents	Last Quarter	Percent	This Quarter	Percent
1. Advertising	1.00 to 2.00	$3,074	1.49	$3,096	1.28
2. Sales	2.00 to 5.00	$8,253	4.00	$9,676	4.00
3. Office Expenses					
Staff	1.50 to 3.50	$6,705	3.25	$7,862	3.25
Rent	0.35 to 1.20	$1,450	0.70	$1,450	0.60
Office equipment	0.10 to 0.50	$144	0.07	$73	0.03
Telephone	0.10 to 0.75	$475	0.23	$508	0.21
Computer*	0.10 to 0.35	$62	0.03	$387	0.16
Office supplies	0.05 to 0.20	$62	0.03	$73	0.03
4. Job Expenses					
Vehicles	0.75 to 2.50	$454	0.22	$484	0.20
Job supervision	1.50 to 2.50	$4,127	2.00	$4,838	2.00
Tools & equipment	0.15 to 0.50	$268	0.13	$169	0.07
Service & callbacks	0.20 to 0.40	$557	0.27	$387	0.16
Mobile telephone	0.05 to 0.30	$165	0.08	$169	0.07
Pagers	0.05 to 0.15	$83	0.04	$73	0.03
5. General Expenses					
Owner's salary	4.00 to 6.00	$10,832	5.25	$12,700	5.25
General insurance	0.25 to 100	$2,579	1.25	$3,048	1.26
O.C.R.A.**	1.00 to 4.00	$4,126	2.00	$4,838	2.00
Interest	1.75 to 3.00	$4,498	2.18	$5,298	2.19
Taxes***	0.00 to 3.00	$928	0.45	$1,161	0.48
Bad debts	0.00 to 0.15	0	0.00	$290	0.12
Licenses & fees	0.10 to 0.25	$62	0.03	$48	0.02
Accounting fees	0.10 to 0.25	$309	0.15	$290	0.12
Legal fees	0.15 to 0.30	$268	0.13	$435	0.18
Education & training	0.10 to 0.20	$144	0.07	$193	0.08
Entertainment	0.05 to 0.10	$619	0.30	$702	0.29
Association fees	0.05 to 0.10	$144	0.07	$145	0.06
TOTALS	15.45% to 38.20%	$50,388	24.42%	$58,393	24.14%

* Computer expense = hardware, software and support services.
** Operating Capital Reserve Account. You may reduce this by 70% to 90% to maintain the account once you've reached the goal for the account.
*** State and local taxes will vary widely with your location and business structure. Federal taxes are not included.

Figure 2-10
Quarterly overhead review form for a typical new home construction company

Quarterly Overhead Review
Specialty Contractor
(Computations based on annual volume built and collected)

Overhead Item	Low / High Percents	Last Quarter	Percent	This Quarter	Percent
1. Advertising	1.00 to 2.00	$1,089	1.84	$1,200	1.86
2. Sales	1.00 to 2.00	$1,184	2.00	$1,291	2.00
3. Office Expenses					
Staff	3.00 to 7.00	$3,847	6.50	$3,789	5.87
Rent	0.35 to 1.20	$475	0.80	$475	0.74
Office equipment	0.10 to 0.50	0	0.00	$168	0.26
Telephone	0.25 to 0.75	$414	0.70	$471	0.73
Computer*	0.10 to 0.35	$171	0.29	$84	0.13
Office supplies	0.05 to 0.20	$65	0.11	$39	0.06
4. Job Expenses					
Vehicles	0.75 to 3.00	$509	0.86	$549	0.85
Job supervision	3.00 to 5.00	$2,811	4.75	$3,066	4.75
Tools & equipment	0.20 to 0.75	$154	0.26	$148	0.23
Service & callbacks	0.10 to 0.50	$65	0.11	$97	0.15
Mobile telephone	0.05 to 0.30	$95	0.16	$123	0.19
Pagers	0.05 to 0.15	$77	0.13	$77	0.12
5. General Expenses					
Owner's salary	6.00 to 8.00	$4,290	7.25	$4,679	7.25
General insurance	0.25 to 1.50	$917	1.55	$994	1.54
O.C.R.A.**	1.00 to 4.00	$1,627	2.75	$1,775	2.75
Interest	0.50 to 0.75	0	0.00	0	0.00
Taxes***	0.00 to 3.00	$426	0.72	$458	0.71
Bad debts	0.00 to 0.30	0	0.00	0	0.00
Licenses & fees	0.10 to 0.25	$130	0.22	0	0.00
Accounting fees	0.15 to 0.30	$112	0.19	$136	0.21
Legal fees	0.15 to 0.40	$130	0.22	$213	0.33
Education & training	0.15 to 0.30	$47	0.08	$84	0.13
Entertainment	0.10 to 0.20	$83	0.14	$110	0.17
Association fees	0.10 to 0.20	0	0.00	$329	0.51
TOTALS	18.50% to 42.90%	$18,718	31.63%	$20,355	31.54%

* Computer expense = hardware, software and support services.
** Operating Capital Reserve Account. You may reduce this by 70% to 90% to maintain the account once you've reached the goal for the account.
*** State and local taxes will vary widely with your location and business structure. Federal taxes are not included.

Figure 2-11
Quarterly overhead review form for a typical specialty contracting company

Chapter 3

Establishing the Correct Markup for Your Company

Let's start a construction company; you can be the president and I'll be your partner. We're going into business for the first time. We'll use a remodeling company as the example because the rate of business failure is higher in residential remodeling than in other areas of construction (although not by much). In my experience, remodeling companies generally have more day-to-day problems to deal with than other construction businesses.

We've done the research for our company and concluded that by working by ourselves, and hiring subcontractors as needed, we can expect to sell, produce, and get paid for a volume of $150,000 in our first year. This number could be anything — it doesn't really make a difference for our purposes. The math is the same, and it's the mathematical procedure that we're going to look at with our company.

Always keep this in mind: Total sales means nothing if you don't get it built, and then collected. Forgetting this subtle item can undermine your best plans if your primary focus is on sales rather than on a balance of sales, production, and collections.

Let's assume that we want to make a minimum net profit of at least 8 percent. Most remodeling contractors would be happy to make 8 percent (many would be happy with any profit at all). And, after calculating our overhead, we project it will be 25 percent of our first year's dollar volume built, or $37,500.

The Basic Formula for Markup

This is what we can project about our company's first year:

- Dollar volume built = $150,000
- Overhead expenses = $37,500
- Net profit at 8 percent = $12,000

Here's how we'll compute our markup:

1. Add overhead expense and net profit:

   $37,500
   +12,000
   $49,500

2. Then subtract that total from our dollar volume to find the job costs:

   $150,000
   − 49,500
   $100,500

Our projected job costs for our first year in business are $100,500. In other words, that's the amount we expect to spend the first year to build the jobs we've projected we'll sell. We can use this formula to compute the markup we need to charge:

Total Dollar Volume Built ÷ Job Costs = Markup

3. Divide $150,000 by $100,500:

   $150,000
   ─────────
   $100,500

4. Markup = 1.49

That's not so hard, is it? But how do we use that number now that we have it? To arrive at the correct sales price for each job that we do, we must first estimate the actual costs for the particular job. Then we'll multiply the estimated job costs by our markup of 1.49. The result is a sales price that will allow us to pay all of our job costs, all of our overhead expenses, and make our projected 8 percent profit. It's just that simple.

Here's the next obvious question: "What if I don't make my projected dollar volume built?" We'll deal with that a little later in the book, but for now let me give you the best answer I can. The difference between successful contractors and the ones who don't survive is that the *successful contractors make sure they do hit their sales and production goals.* They don't allow anything to distract them from their goal. In short, they make it happen. If they get sick, they get help. If they get hurt, they get more help. If they get lazy— well, sorry, that's not something that this book can fix. Plain and simple: *you must make it happen.*

Let's Do Some Sample Problems

Now that we have a formula for the correct markup for our company, let's see how it works. Let's say we just estimated a small remodeling job for Joe and Mary Smith and our total job costs are $1,750. Our sales price, using the markup that we established for our company, should be $2,608.

Let's review the math:

$1,750 job costs
× 1.49 markup
───────────────
$2,608 sales price

This markup percentage should remain the same for all the jobs that we sell, regardless of size. I can see you're about to protest, but don't jump too far ahead. Stay with me here. We'll address the issue of very large jobs versus very small jobs in a bit. For now, just stick with what we're doing and get a good understanding of the process.

A quick check using the following formula will tell us if we're using the right markup.

Markup × Job Costs = Total Volume Built

$100,500
× 1.49
─────────
$149,745

That's close enough to the $150,000 projected dollar volume to show us that this markup is correct.

There are sample problems at the end of the book that you can use to demonstrate that this markup theory really works. I encourage you to work all the problems. They'll not only help make you comfortable with the math, but they'll establish the thinking process you'll need when a new situation in your business causes you to consider changing your markup. The following examples also provide you with some practical applications of how this theory works.

Residential Remodeling

For our first example, let's look at several situations you'll face almost every day in a remodeling company. We're going to calculate the markup and sales price of a proposed remodeling job, and some related numbers to give you some practice in developing your own numbers.

Let's assume we've been out to see the Jones family: Mr. and Mrs. Jones and their three children. They've run out of room in their 1,400-square-foot home. We've designed a plan for them that includes a new master bedroom and bath, and some changes to the interior walls to convert the existing bedrooms into three larger, workable bedrooms. The changes also involve reworking the existing bathroom and adding a hallway.

We've done our homework; we know the home inside and out. We have firm written quotes from all our specialty contractors, quotes from suppliers for the various materials we'll need, and we've come up with a final job cost of $28,875. These numbers are solid — right on the button.

Last November, we calculated our overhead for this year, and set our overhead budget at $87,950. To date we're exactly on budget. Our sales goal for this year is $287,500 and we're making our projections happen. We're right on schedule for the number of jobs sold, built and collected. We want to make a profit of 10 percent this year.

Here's what we need to know:

1. What is our percentage of projected overhead for the year?
2. What is our projected profit in dollars for the year?
3. What are our projected job costs for the year?
4. What is our current markup?
5. What should the sales price be for the Jones job, including markup?

Before you read farther, why not grab your calculator and see if you can answer these questions. If you've followed along to this point, you know that you have all the information that you need to find the answers to each of the questions above. This is the kind of effort you need to expend if you're going to make it in this business.

If you come up with the wrong answers, don't worry — the world won't end. Just figure out what you did wrong and try it again. Once you get used to applying these formulas, you won't make mistakes when you're calculating the sales price of your real jobs in front of your customers. That's what practice is all about.

Let's work through these problems.

1. What is our percentage of projected overhead for the year?
   Expense ÷ Projected Sales = Overhead Percentage
   $87,950 ÷ $287,500 = .3059 or 31%

2. What is our projected profit in dollars for the year?
   Projected Sales × 10% = Profit
   $287,500 × .10 = $28,750

3. What are our projected job costs for the year?
   Total Sales − (Overhead Expense + Profit) = Job Costs
   $287,500 − ($87,950 + $28,750) = Job Costs
   $287,500 − $116,700 = $170,800

4. What is our current markup?
   Total Sales ÷ Job Costs = Markup
   $287,500 ÷ $170,800 = 1.6832 or 1.69

Note: While rounding this to 1.68 would be technically correct, it's a good idea to always round up when figuring your markup. Dropping that .0032 may not seem like a big loss, but over time it will add up. Raising your markup by 1 percent won't make any difference in your ability to sell this job, but it will make a big difference in your bottom line over the year.

5. What will be our sales price for the Jones job, including markup?
   Job Cost × Markup = Sales Price
   $28,875 × 1.69 = $48,798.75 or $48,799

I can hear the cries of anguish from the large majority of the readers, even as I write this book. "Nobody can charge 1.69, you'll never get the work. Highway robbery! Tar and feather that crook!"

If those or similar thoughts are going through your head right now, you're going to have a tough time surviving in this business. Please answer just one question for me. Where's the mistake in the math? Show me what's wrong with the numbers we've just worked out. You've seen how we've arrived at the markup for our company. The math is there, and it's correct. That profit is more than reasonable for all the time and effort that'll go into the jobs our company will build this year. Our $28,750 profit for a year's work is less than most government employees earn. Most of them don't work nearly as hard as we do, and they take weeks of paid vacation and holiday time.

Sure, this is a hypothetical problem for a hypothetical company. But where would you cut to lower your markup? You might say that you could cut the overhead percentage down. Show me where. The 30.5 percent overhead figure that we're using is probably a lot closer to reality for most remodeling contractors than some arbitrary number under 20 percent that so many contractors think they must use. So where do you cut?

Let me ask you one more question. If someone hires you to do a job for them, don't you do the best job you can for them? Don't you research and complete the work to the best of your ability? Well, that's how I approach things, too. And that's how I came up with these numbers.

The bottom line is this: if you don't know how to arrive at a markup figure, or choose to ignore the knowledge you do have, you're well on your way to going broke. Knowing this, if you don't make an immediate adjustment to the way you're doing business, your business will soon be history. I'd like to be the first to wish you well in your next endeavor.

You must trust your math! If the math dictates a 1.69 markup but you chose to cut your sales price, how will you pay your expenses? They won't just go away.

New Home Construction

A new home contractor should be able to produce at least 1½ to 2 times the volume that a remodeling contractor produces with the same number of people. I've talked with many contractors who can produce double what the

Chapter 3: Establishing the Correct Markup for Your Company

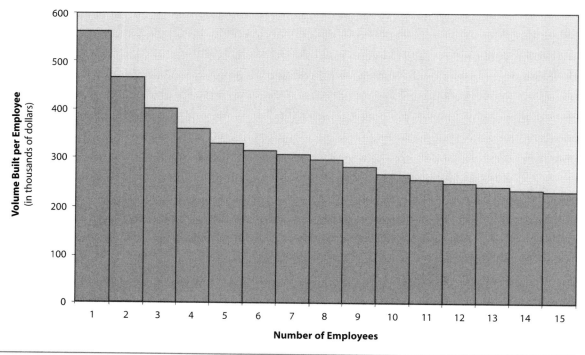

Figure 3-1
Volume built per employee in new construction

average remodeling contractor will produce, and others who claim to triple that amount. Again, the size of your jobs and how effectively you use your employees and/or your subcontractors will determine the numbers. It's a simple matter of organization, automation and discipline. The more you have of each, the more you can get built, and coincidentally, the more money you make!

If you study the graph in Figure 3-1, it'll give you a visual yardstick for measuring what your production should be based on your number of employees if you're in the new home construction business. As with the graph shown earlier for remodeling businesses, the production curve tends to flatten out as the number of employees increases. Your production should be within 5 percent of that shown on the graph. If it's less, you should consider getting some outside advice on how to increase your per-employee production.

Let's look at some average numbers for a new company in new home construction. You'll notice that we calculate markup for new home construction exactly the same way we did for remodeling.

First let's calculate our markup. We've figured our overhead for our first year in business, and project it will be $87,950. Of course it could be any number, larger or smaller, but I've kept it the same as the remodeling company for reasons you'll soon see. Keep in mind we're after the process here, not the number.

We have decided to build four homes this year. We will build two of our #1310 models at a cost of $115,000 each (no land expense) and two #1525 models at a cost of $138,000 each. We have set a goal of making $55,000 net profit. What should our markup be to arrive at the sales price for these homes?

Before you read on, why don't you see if you can take this information and calculate the correct markup for the company? This exercise will tell you if you have a grasp on the ideas that we've covered up to this point.

This is what we've projected for our first year:

- Job Costs = $506,000
- Overhead Expenses = $87,950
- Profit = $55,000

To compute our markup:

1. Add Job Costs, Overhead Expenses and Profit to get the Total Volume Built:
$$\begin{array}{r} \$506,000 \\ 87,950 \\ +\ 55,000 \\ \hline \$648,950 \end{array}$$

2. Divide the Total Volume Built by Job Costs to arrive at Markup:
$648,950 \div \$506,000 = 1.2825$, rounded to 1.29

You can check your figures by using our formula:

Markup × Job Costs = Total Volume Built
$1.29 \times \$506,000 = \$652,740$

Looking at the numbers we've just calculated brings us to some very interesting points. As your Total Dollar Volume Built increases, your percentage of Overhead Expenses is proportionally smaller. When the Total Dollar Volume Built was $287,500 for our remodeling company, the overhead was 30.5 percent of that amount. Now our overhead is only 13.6 percent of our Total Dollar Volume Built. And, as you can see, your markup also goes down proportionally as well.

Combining New Home Building and Remodeling

I want to say something here that's a little off the subject. I believe few if any general contractors can both build new homes and do remodeling successfully at the same time. Sure, many general contractors try, but I've never met one that has successfully done it with a consistent, long-term

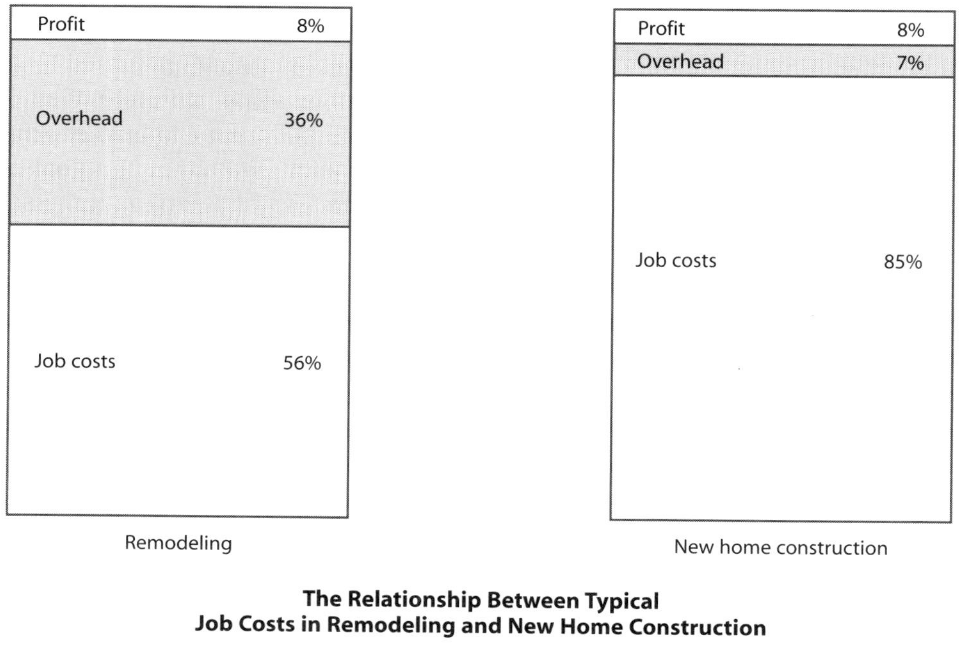

Chapter 3: Establishing the Correct Markup for Your Company

```
Remodeling:
  Profit     8%
  Overhead  36%
  Job costs 56%

New home construction:
  Profit     8%
  Overhead   7%
  Job costs 85%
```

**The Relationship Between Typical
Job Costs in Remodeling and New Home Construction**

Figure 3-2
Job costs and overhead in remodeling and new home construction

profit. *New home construction is a commodity business; remodeling is a service business.* They're as different as daylight and dark.

I and a number of other very good contractors have tried to do both, and found that mixing these two types of construction just doesn't work. Not only are the economics of the two businesses very different, but so is the physical work itself. As a rule, the same crews can't do both types of work. My advice is to focus on one or the other, but don't switch back and forth.

If you do get into any combination of residential remodeling or new construction, or specialty or commercial work, your sales volume for one division will most certainly be different than the other division. It follows that if your volumes are different, your markup must also be different. That's why you can't use the same markup for remodeling that you use for new home construction. It's also one of the reasons that few if any contractors can successfully work in two areas of construction at the same time. The graph in Figure 3-2 illustrates what I'm talking about.

Having said all that, here's a final word of caution on this matter. If you've developed your markup as outlined and your numbers are based on building new homes, you'll lose money if you use that same markup for a remodeling job. Do it consistently and you'll go bankrupt. This of course assumes that your volume of business in new home construction is larger than the volume of work that you do in remodeling.

The opposite, however, works very well. If you use the same markup for new construction that you've been using for remodeling, you'll make money. (If you can get any jobs at that price, that is.)

And here's some more advice. You must keep separate books for each type of construction that your company is involved in, one for new homes and one for remodeling work. This is mandatory. If you lump everything into one set of books, you'll never be able to figure out where you are in assigning expenses. And if you can't figure out where the expense should go, you certainly won't be able to compute an accurate markup for each different job. This is one of the main reasons that I don't recommend trying to do both remodeling and new construction work at the same time. Contractors are notorious for their bookwork. Few, if any, will actually take the time or pay the expense to keep two separate sets of books. And that's enough said about that!

The 10 Percent Myth

Now before we go on, let's once and for all put to bed the myth that "you can only get 10 percent on your costs and still sell a new home."

If we take the $506,000 job costs for the four homes we're going to build and only add 10 percent ($50,600) like the all-wise and all-knowing experts say we should, we'll end up with a Total Volume Sold of $556,600. That means we'll have just $50,600 to cover all of our overhead expenses (which we know are $87,950) and our profit (what profit?). Hmmmm . . . Very interesting. Unless we have a new form of math in this business, it looks like we're $37,350 in the red and we haven't even started building! Clearly, that just won't work!

Here's the reality of those numbers. First, it's clear that there's no possibility of making a profit. As I've already explained, profit is not, nor should it ever be considered to be, the owner's salary. Profit is profit, and the owner's salary is part of the overhead expenses. They're two separate things. If you'll accept that premise, you're on the road to making some money in this business.

If you believe that paying yourself for the work that you do is enough, and you don't need to make a profit too, then why suffer through all the headaches of being in business? You can go work for that other guy, put in your 8 to 10 hours a day and go home and do your own thing. There'll be no bookwork to do when you get home at night, no phone calls to make, no equipment to fix, no ungrateful customers to deal with, no money to collect, and most importantly, no bills to try and figure out how to pay with money that's not there. Life would be simple!

OK, maybe your overhead isn't as high as the 1.29 figure that we've used, or maybe you don't mind working for little or no profit. That's your choice, and it'll certainly lower the number for your markup. The question is, how long are you going to work that way? If you're in business, and you're not making a profit or even bringing enough home to pay your bills (which is the case with at least 30 percent of all contractors at any given point in time), your company isn't going to make it.

"You must run your business as a business. Your main reason for being in business isn't to provide jobs for other people. You're in business to provide a service to your customers and to make a profit doing it."

If you're only building four homes a year, your markup is most assuredly going to be higher than the companies who build 75 or 100 or even more homes each year. If you do the numbers for some of the really big companies, those building 800 or more homes each year, their markup might be as low as 1.0875. As a small businessperson, you simply can't compare your markup to theirs. It just doesn't work! But I can guarantee you that they didn't start off in business charging a small markup. They got that big because they've always known exactly what their numbers were and they made darned sure they got those numbers on each and every home that they built. They always make a profit. Companies at this level have better sense than to build a job at a loss, for any reason.

A Business Is a Business

Now let me cover another argument that seems to come up over and over again when I talk to contractors. It's the old song, "I have to keep my guys working, or I'll lose them." Or, other verses from the same song: "It keeps the money flowing through my company so I can keep my bills paid," and "It gives me something to do until a good job comes along that I can make money on." There are more verses as well, but these are fairly representative of the thinking (or lack of thinking) that I run across.

You must run your business as a business. Your main reason for being in business isn't to provide jobs for other people. You're in business to provide a service to your customers and to make a profit doing it. If there are no profitable jobs to work on, then everyone goes home until there are. That's the nature of construction. If you've been in construction for more than a day, you'll have heard that there are some down times. We all have them; they're part of the business. If you or your employees can't handle that, then you'd all better go look into another profession, and quickly. You're in the wrong business! So, if your choice is taking a job at a loss, or sending your people home until you get a job that you can make a profit on — then everyone goes home.

Chapter 3: Establishing the Correct Markup for Your Company

For those of you who still don't get it, here's something to ponder. Do you think that any of your workers would be willing to give you back part of their wages after the job costing has been done on a particular job and you can show them that you lost $37,350? Don't hold your breath while you wait for that to happen! If good jobs are slow in coming and some of your crew decides to move on, there'll be others who'll take their places. Sure it's inconvenient, but your company will continue, and you will survive.

Taking a job just to keep money flowing through a company is just as foolish as taking a job to keep your crew working. When you get the job done, you'll have more bills to pay than you'll have money to pay them. From that point on, you'll be playing catch up. If you take a job that lands you in the red, at some point you'll have to charge full markup *and more* on a whole series of jobs until you can get your company back on track. And, if you don't, your company will fail. Why put yourself through that misery? We'll talk some more about this later in the chapter.

It's important for you to get a good handle on the ideas that we've covered up to this point. If you don't feel comfortable with them, stop here and take a break. Give it a day or two to sink in, and come back to it after you've allowed your mind to work it over for a while.

Specialty Contractors

Some of you may be surprised to learn that the percentage for overhead expenses in the specialty building trades normally won't fluctuate more than plus or minus 4 percent from general contractors' overhead. There's a commonly-held belief that specialty contracting is somehow "different," and so the overhead numbers must also be different; for most people that means much, much lower.

If you're a specialty contractor, take the time to develop a list of your overhead expenses like the one I did for Figure 2-6 on page 37. It's a good exercise that should also give you more confidence in establishing a sales price for your work.

There are two major differences in the overhead expenses for specialty and general contractors. First, specialty contractors normally don't have the sales expense that general contractors have. That's assuming, of course, that they've done a good job of letting the general contractors who can use their services know that they're in business. If they've done that, then good "word-of-mouth" advertising resulting from their work will get them referrals and repeat work from general contractors on a regular basis. That's also assuming that the specialty contractor chooses to work primarily with general contractors, as opposed to soliciting and doing work for the general public. If a specialty contractor does do some or all of their work for the public, they'll have the same advertising expense as a general contractor. Working only for general contractors and not advertising lets specialty contractors lower their overhead expense by 6 to 12 percent.

Chapter 3: Establishing the Correct Markup for Your Company

The second reason that specialty contractors' overhead may be lower is because they do a smaller volume of business (or their job size is smaller, however you want to put it). Their overall cost of doing business is less, in terms of dollar volume and percentage of expense. That's a function of spending less money on variable overhead items.

Some specialty contractors have an average job size of only $550 to $1,200. Because of smaller job size and lower overhead, they'll normally have a markup that's lower than a remodeling contractor doing the same volume of work, but higher than contractor doing new home construction or commercial work.

As a specialty contractor, once you have your overhead costs "nailed down," you can begin to develop a plan to incorporate those costs into the sales price of your jobs. It works just like the plan for remodeling jobs in the section on residential remodeling.

I didn't include a chart showing the Volume Built Per Employee for specialty contractors for a couple of reasons. First, there are just too many different kinds of specialists out there. Each would need a chart that's a little different. Second, you can get a good idea of about where you should be by looking at the chart for remodeling contractors, Figure 2-8 on page 43. If the majority of your work is for general contractors, then expect 5 to 8 percent less volume built per employee. If the majority of your work is for home or building owners, your volume built per employee should be very close to what you'll find in Figure 2-8.

Keeping a Schedule

There are some problems in this business that affect specialty contractors a little more than any other group. The first involves setting a schedule and keeping to it. In construction today, the general or specialty contractor who can set a schedule and do all their work *on time* is rare. That's one of the complaints I hear most often.

When you set a schedule for the work that you do, *make that schedule happen*. If you're working for a general, tell them up front that you'll be calling one week prior to your start date and again the day before the start date to check the job progress. Assure them that you'll be on the job, on time, and you'll get the job done by the time and date you've arranged. *Now you've forced the general contractor to get that job ready for you*. He's on schedule, you're on schedule, the job moves smoothly, and everyone is happy.

Who do you think that contractor's going to call for his next job? If you follow though, make the phone calls, show up on time, and do a good clean job, you'll get most (or all) of that general contractor's work in the future. And here's the best part. If you do all those things, you can charge from 10 to 20 percent more for your work. You're worth it. Your markup can easily go from 1.35 or 1.40 to 1.45 or higher.

Advise the general contractor that if your crew arrives at the job and the job isn't ready, you'll bill them for trip time at $55 an hour (or some other amount that'll get their attention). Make sure they understand that you'll carry through on this billing. Convey it politely, but firmly. If they don't want to pay that charge, they'd better have the job ready for you or notify you in plenty of time that it won't be ready. Let them know you need at least two full days notice.

If the general contractor doesn't have the job ready, be sure you follow through and bill them for the travel time to the job. Don't let them off the hook. You have to spend your time or pay your crew to go to the job site. Why should you pay for someone else's lack of organization? And when they call and tell you that the job is now ready for you, fit them back into your schedule *when it's convenient for you to do the job*. Don't plug them back into your schedule in front of other jobs that you've previously scheduled, and who are operating on time. If you do, you'll be compounding the problem. Don't reward tardiness or disorganization by inconveniencing yourself, your employees or other general contractors.

Low Bids

Another problem that affects specialty trades more than others is competitive bidding. Many specialists spend a lot of time and money learning this the hard way. As a subcontractor, you'll be asked by general contractors to come out and give them a "bid" on some work. If you want to do this as a means of developing new business, then go after it. But here's a word of caution for the sub and some good advice to the general contractor on the subject of competitive bids.

If you're a specialty contractor who's given three estimates to a general contractor who hasn't given you any work, it's time for a talk with that contractor. You need to take a stand at some point. Your goal is to secure work for your company, not allow yourself to be used just to help the general develop numbers for "bids" for his or her customers. And let me add a subtle, but important, point here. Base your continued work with each general contractor on the total dollar volume of work that comes your way, not the number of jobs.

There are general contractors out there who only care about the low bid. They think your sole function is to supply them bids so they can sell their jobs as low as possible. They'll cut your profit to nothing so they can make a little more money. Generals looking only for low bids are usually the ones who're going broke — or are already broke but don't know it yet. I suggest that you pass on doing business with them. Look for general contractors who are fair and forthright.

Generals, I hope you're listening here. This applies as much to you as it does the specialty contractors who work for you. Don't use subs to bail your fat out of the fire. If you ask a sub for a price on a job and you get it, award that work to the company that helped you put that job together.

If you're a specialty contractor and you've given a quote to a general, they should notify you about the job in a reasonable amount of time. Being in business is tough enough without waiting a couple of weeks to hear about a job that you thought you were going to get, only to discover that it went to someone else. We've all had that happen to us. If you've given someone a bid and you don't hear from them, give them a call. You can't operate on speculation. If you have the job, push for your start date. You need to keep your business on a schedule — and the general contractor needs to get organized. Whether you're a specialty contractor or a general contractor, treat everyone with the respect that you'd like to get. Both your time and their time are valuable.

Whether you're a specialty contractor or a general contractor, you're under no obligation to give out bids to anyone. Always try to find out how many bids your potential customer or general contractor is getting. If I find out a potential customer of Stone Construction Services has asked for quotes from more than two other contractors about a particular job, I pass on that job. Shopping bids is a sure indication that price is the customer's first (or only) consideration. If the homeowner or general is more concerned about the price than the end result, then I know from experience that I'll have problems with them during construction. It's best not to get involved with jobs like that.

If a customer's sole criterion for picking a sub is price, consider that any time spent working up your bid may be time wasted. Factor that time into your bid price, even though you know there are companies out there willing to give their work away. If you do get the bid, you'll have the cost of your time covered, at least on that one job.

Commercial Work

Contrary to what most commercial contractors will tell you, you should use the same approach in establishing your sales price for commercial work as for remodeling, new home construction or specialty work. The math doesn't change. Commercial work normally has a much higher dollar volume in any given year than other areas of construction, so your markup is normally lower than even for residential construction.

Volume Built per Employee for Commercial Work

I don't have a chart for the Volume Built per Employee for commercial work. After talking to dozens of commercial contractors and doing considerable research, I still haven't come up with any average numbers you can accurately use to project the volume of work per employee. That's because commercial jobs span a much wider range of volume than you find in other areas of construction.

For example, I know one commercial contractor who does mostly tenant improvement work. He did about $4.5 million last year in jobs sold, built and collected. He has a dozen full-time employees, and his average volume built per employee is around $375,000. Another commercial contractor builds large concrete tilt-up warehouse-type buildings. His volume built was $58 million last year. With 22 employees, his average volume built per employee was around $2,636,000. They're both working successfully in commercial construction, but as you can see, their numbers are very different.

If you're not achieving your goal of 8 percent net profit, then you need to figure out your own volume per employee, and try to improve on it. Call a conference of your key people, as we discussed in Chapter 2 under Increasing Production, and tell them that you want to see an increase in your numbers in the next 90 to 180 days. Set some number as a goal, probably about $25,000 per employee to get started. If you're on top of what you're doing, that should be an easy number to reach. How much higher you want to go in pursuing additional increases is up to you. I think I'd try to increase your volume per employee by about $25,000 every four to six months until you've reached a plateau. Then give it a rest for four to six months. However, you must be diligent in watching production levels. Once people get used to what they're doing, they tend to become comfortable, and begin slacking off. That's normal. When that happens, get back after it again. You and your crews should never be comfortable too long — you should always strive to do better!

Sliding Scale Markup

There's only one exception I can think of to using a fixed markup based on a percentage of your total volume built. That exception is using a sliding scale markup. You can use a sliding scale to lower the markup on your larger jobs as long as you have the discipline to increase the markup on the equivalent dollar amount of smaller jobs. You have to make up the difference in total dollars sold, built and collected brought into your company by way of contract. Adjust your markup as you will, but at the end of the year you must be able to divide your total sales by your total job costs and get the amount you projected at the beginning of the year. If the number you arrive at is the same or larger than your projected markup, you've mastered the sliding scale approach and you'll be making money. The best test, however, is whether or not you've made a minimum of 8 percent net profit!

I don't advocate using a sliding scale markup. It's too easy to think that you'll cut your markup a little on one job to ensure you'll get it, and then make it up on the next three jobs. Then the next job comes along and you feel you just can't increase your markup on that one . . . and the next one, and the next. Pretty soon the year has passed and you're still playing catch-up.

I've seen a lot of companies try using a sliding scale markup over the years. Frankly, it seldom works. In 95 percent of the cases I've studied, when the company does raise their markup on smaller jobs to compensate for lowering it on larger ones, the higher markup generates a price so high that it puts them out of the running for most of the jobs they go after. They'll coast along for a while using the sliding scale markup and relying on their company name and reputation to continue to get them jobs. Eventually, however, the tag of "high priced" will catch up to them and their sales will take a nose-dive. In a panic, the owner will cut their markup so they can charge a reasonable amount and bring back their sales, but usually it's too late. The damage has been done. It's very hard to recover and regain your market position once you've been labeled "high priced." Will they make their 8 percent profit in the end? No. Will their business survive the scramble? Some do, but many don't.

A Sample Company

Now, let's look at how a sample general construction company might use a sliding scale markup. We'll call them Coffin Hoist Construction. The company has an overhead of $135,800 for the year. On average, across the board, the contractor estimates he'll sell about $490,000 and he wants to earn a minimum of 8 percent net profit. The math shows that he should use a 1.5549 or 1.56 markup.

The first year or two that CHC is in business, they'll have to estimate the cost range and the markup assigned to each range. With careful monitoring, however, by the end of their third year CHC will have actual numbers to use. They'll have built enough jobs to be able to group their average job sizes into six categories. Then they can set a markup for each size category, with a difference of about 3 to 5 percent, but it can be as high as 10 percent, between each category's markup.

Using the average job size for the year as the norm, they can assign their normal markup to that group. Then they'll adjust their markup up or down depending on the job costs of each of the other groups. The chart in Figure 3-3 illustrates this for Coffin Hoist Construction. Of course they'll have to adjust their markup from time to time as their overhead costs change.

If you use the sliding scale markup, check your numbers at least quarterly to be sure that your overall markup number is meeting at least the minimum markup you've set for the company. This is no place to get lazy. It takes careful and continuous monitoring of the sales price of jobs to keep things in balance. The bottom line is that you must maintain the percentage of gross profit necessary to cover all overhead expenses and make your predetermined profit. If you don't, you're history.

Figure 3-3
Sliding scale markup calculations

Coffin Hoist Construction Sliding Scale Markup Calculations						
Average Job Cost	Sliding Markup	Sales Price	Number of Jobs	Sales per Markup	Cumulative Sales for Year
$2,200	1.75	$3,850	7	$26,950	$26,950
$5,075	1.70	$8,628	5	$43,140	$70,090
$11,640	1.65	$19,206	4	$76,824	$146,914
$20,470	1.60	$32,752	3	$98,256	$245,170
$42,700	1.50	$64,050	2	$128,100	$373,270
$80,500	1.45	$116,725	1	$116,725	$489,995

Let's look at how CHC's figures came out. They wanted to make a minimum of 8 percent net profit on $490,000 volume of work for the year, or $39,200. If you take the total sales as shown for the year of $489,995 and multiply it by 8 percent, you get $39,199.60. You can't do much better than that! As a matter of fact, if you can get to within .04 percent of your figure for actual jobs sold, built and collected, you'll be doing very well.

Our proposed markup was 1.56, but our average markup ended up to be closer to 1.66 over the year in order to reach our goal. We had to sell 19 jobs over our proposed markup in order to balance out undercutting our markup on our three large jobs. That's the downside to this approach. Few contractors have the discipline to maintain the higher markup that they need on their smaller jobs, or to monitor their sales on an ongoing basis. Believe me, it's far easier to use one markup and go sell another job or two, than spend the time wrestling with your numbers for a sliding scale.

A Review of Volume, Overhead and Markup

We've established what the minimum items of overhead are for most companies. We also know that overhead, depending on whether you're in business as a general or a specialty contractor, can range from a minimum of 15.45 percent to 53.00 percent in some cases. Now let's review some applications of markup to highlight the various shades of the markup theory.

There's an old saying that flits about the construction industry like the ghosts in Figure 3-4. It goes something like this, "As your volume goes up, your overhead expense comes down!" Maybe you've heard it? Most of us who've been around a while have heard at least one contractor say, "Well, that job was quite a bit bigger than the jobs I normally do, so I was able to charge less since my overhead was lower." I sincerely hope that you've never said anything like that. If you have, watch out.

Chapter 3: Establishing the Correct Markup for Your Company

Figure 3-4
Construction industry ghosts

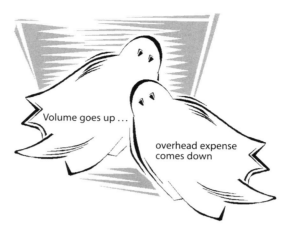

The Relationship of Overhead to Volume

Here's a dose of reality: As the dollar volume of jobs sold and built goes up, your overhead expenses go up as well. They do not go down. Let me repeat this for you. *As the dollar volume of jobs sold and built goes up, your overhead expenses also go up, not down.* If you will take a minute to look back at Figure 2-7 on page 38, which shows the comparison of overhead to dollar volume built, you'll see that this is true.

As you build your jobs, your dollars spent for overhead go up. The *percentage* of your overall budget that you spend on overhead does come down, but only in proportion to the total volume of work being done. This is a perfect example of how that old ghost of a saying has been perpetuated down through the years. Few have ever bothered to think it through. There's some truth to it, but only when you're dealing in *percentages of your total volume built.*

Don't Compare Apples to Oranges

Contractors doing remodeling or specialty work are in a service-oriented business that's labor intensive. As they do more work and their sales volume goes up, they acquire more overhead expenses, both fixed and variable. Contractors doing either new home or commercial building construction are in a more commodity-driven business, and they have lower overhead expenses. Both, however, operate under the same mathematical rules that govern the total overhead and the application of the proper markup.

Don't try to compare what you're doing in construction with what the local supermarket or discount stores are doing. They're service companies too, but they're selling lots of merchandise and their percentage of overhead is much, much smaller than yours. The VP of a large supermarket chain here in the Northwest told me that his company's goal is to make a .75 percent net profit in their grocery department. Now considering that they have over 75 stores, each selling at least $750,000 in groceries every day, seven days a week, you can see that we in construction would be hard pressed to compare our percentages to theirs.

If you have doubts about the relationship of volume and overhead, check your own company books for any two-year period when the volume of the second year was more than the preceding year. If your books are in order, you'll see that your true variable overhead expenses went up, but your percentage of overhead (fixed and variable, based on the total sales price of your jobs) went down.

And What About Markup?

I can hear you saying to yourself, "If my expenses for overhead go up, shouldn't my markup go up as well?" The answer to that question is no. This is another common point of confusion.

After your second year in business, you should be able to hold your overhead expenses to within 1 to 2 percent of your budgeted amounts. If you can't, it's normally because you didn't stick to your budget. So as your volume goes up, your fixed overhead expense should remain close to your budgeted numbers for the year. That means your markup, as a percentage of the relationship of those numbers, stays the same.

Budgeting Increases into Your Overhead

Spending more for overhead items is OK, providing you go out and sell enough additional business to compensate for the additional overhead expense. You need to determine how that expense item is going to impact your overall numbers *before* you add it into your budget.

Contractors are constantly spending money on things that aren't in their budgets — often because they don't have a budget to begin with. Or, if they do have a budget, they don't know how to adjust for the unplanned expense by selling additional volume. What you need is a formula that will tell you exactly what additional dollar volume of jobs you must sell, build and collect to give you the additional money that you'll need to make an unbudgeted expense.

So do you have a formula like that? When you want to buy something that's not in your budget, can you quickly calculate exactly what additional business you'll need to sell to compensate for the expense? If not, let's figure it out. Right now, see if you can come up with a formula that will work. It's not a difficult problem, and should only take a few minutes.

Computing a Formula for Assimilating an Unbudgeted Overhead Expense

Have you successfully arrived at your formula? Let's look at the relatively simple math process you should use when you deviate from your budget. As we go through this process, we'll build the formula.

Unbudgeted expenses happen. That's a given in this business. No matter how much I like to rant and rave about how you must stick to your budget once it's completed, the reality is that things come up that create unbudgeted expenses.

What constitutes an unbudgeted overhead expense? You may find that you need to purchase something for a particular job but you're sure you'll be using it for other jobs as well. If the item isn't a really high-cost item, just add it into your estimate for that particular job. An example might be a set of suction cups for moving large pieces of glass that you need for a sunroom addition. But don't include something as a job cost unless it's a true job cost. Trying to put a computer into the budget for a particular job is stretching things a bit. Computers are definitely an overhead item.

Let's look at two typical cases that might come up during an average year in your business, and consider them from three different approaches. First, we'll look at the necessary increase in sales to compensate for the added overhead expense. Then, we'll look at keeping the total sales the same, but increasing your markup to cover the additional overhead expense. And last, and this approach I never recommend, we'll look at lowering your profit percentage to cover the additional overhead expense.

Using an Increase in Sales Volume — Once again, it isn't the numbers that count here, it's the process that you go through to ensure that you've accounted for all your expenses that's important.

We'll use the following numbers for our examples:

1. Projected sales volume from 28 jobs at $9,600 is $268,000.
2. Projected overhead expense, $6,185 × 12 months, is $74,220.
3. Projected net profit is 10% or $26,800.
4. Projected job costs:
   Total Sales − Overhead and Profit
   $268,000 − ($74,220 + $26,800)
   $268,000 − $101,020
   Job costs = $166,980
5. Projected markup:
   Sales Volume ÷ Job Costs
   $268,000 ÷ $166,980
   Markup = 1.60498 or 1.61

Half-way through the year, you park your truck on the street in front of your favorite lunch spot and an uninsured driver runs into it. The truck, after the dust and smoke clears, is history. For the purposes of this example, we'll assume your insurance won't cover the cost of a replacement truck.

Chapter 3: Establishing the Correct Markup for Your Company

You go shopping and find a good two-year-old vehicle with a price tag of $18,500. There's no question about it, you need that truck. Your credit is good, so your neighborhood banker will lend you the money for the new truck. The actual cost of the truck, including license fees and interest on the loan, will be $21,749. Your business is doing well so you decide that you want to pay off the truck by the end of the year. How much additional business must you sell, build and collect, and what should your new markup for your business be if you buy the truck?

1. Projected sales volume for year is $268,000.
2. Projected overhead expense for year is $74,220.
3. Overhead is what percent of total sales?
   $74,220 ÷ $268,000 = 27.7%
4. New overhead expense for year:
   $74,220 + $21,749 = $95,969
5. New projected sales volume necessary to cover new expense:
   New Overhead Expense ÷ Overhead Percent
   $95,969 ÷ 27.7%
   New sales volume = $346,459
6. New projected net profit for year:
   10% × $346,459 = $34,646
7. New projected job costs:
   Total Sales − Overhead and Profit
   $346,459 − ($95,969 + $34,646)
   $346,459 − $130,615
   New job costs = $215,844
8. New markup:
   Sales ÷ Job Costs
   $346,459 ÷ $215,844
   Markup = 1.6051 or 1.61

Please note that your markup remains the same!

The key here is that you must now sell, build and collect an additional $78,459 of business ($346,459 − $268,000 = $78,459) during the remainder of the year to compensate for the new truck expense. This assumes that all your other overhead expenses will remain the same. Your variable overhead expenses will go up as your volume goes up, but using your original overhead percentage figures should compensate for this increase. If you decide that this large an increase in sales volume is too much, you can spread the expense out over a longer period. If you spread it out over a two- or three-year period, then you simply divide the total amount of the new purchase by 2 or 3 and use that number for your additional expense and include the balance in your overhead budget for the coming years.

Let's look at the math for carrying the truck expense out over a two-year period. First divide the cost of the truck in half ($21,749 ÷ 2 = $10, 875) to get the amount you want to budget into each year's overhead. Now go back to step # 4 in our formula and follow through with the new amount.

4. New overhead expense for year:
   $74,220 + $10,875 = $85,095

5. New projected sales volume necessary to cover new expense:
   New Overhead Expense ÷ Overhead Percent
   $85,095 ÷ 27.7%
   New sales volume = $307,202

6. New projected net profit for year:
   10% × $307,202 = $30,720

7. New projected job costs:
   Total Sales − Overhead and Profit
   $307,202 − ($85,095 + $30,720)
   $307,202 − $115,815
   New job costs = $191,387

8. New markup:
   Sales ÷ Job Costs
   $307,202 ÷ $191,387
   Markup = 1.6051 or 1.61

Again, notice that your markup remains the same.

A bit of a side note here. The construction business is cyclical. You've probably heard this even if you haven't been around long enough to see it. So take that into consideration. If you commit to the new overhead expense while your business is doing well, you should try to absorb as much of the cost into the current year's overhead as you can. But if you're just getting your business up and going, or if business hasn't been growing in the volume you expected, you'll probably want to spread a large purchase over a longer period. The important thing is to work each new purchase into your budget, then stick to your budget. Don't get lazy and let these things slide. I've seen way too many contractors fail because they haven't done their math and accounted for unanticipated expenditures. You have to be disciplined to stay in business.

Using an Increase in Markup — Now let's look at changing our markup to compensate for this additional overhead expense, instead of increasing sales:

1. Projected sales volume for year is $268,000.

2. Projected overhead expense for year is $74,220.

3. Overhead is what percent of total sales?
    $74,220 ÷ $268,000 = 27.7%

4. New overhead expense for year:
    $74,220 + $21,749 = $95,969

5. Projected net profit for year:
    10% × $268,000 = $26,800

6. New projected job costs:
    Total Sales − Overhead and Profit
    $268,000 − ($95,969 + $26,800)
    $268,000 − $122,769
    New job costs = $145,231

7. New markup:
    Sales ÷ Job Costs
    $268,000 ÷ $145,231
    New markup = 1.8453 or 1.85

You've got to increase your markup from 1.61 to 1.85 for the balance of the year to compensate for the additional overhead expense. But understand this: increasing overhead without an increase in sales volume results in less money available to build your jobs. (Job costs are reduced from $166,980 to $145,231.) That means you'll have to be at least 13.03 percent more efficient in everything you do on your jobs.

Can this be done? With the small volume we're talking about in this example and with some diligent efforts on the part of the contractor, it can be done — but it's highly unlikely. If, on the other hand, we were talking about $1.5 million to $2 million in volume and we had to increase our efficiency by 13 percent, we'd have a definite problem.

My point is that you'd probably find it much easier to go out and sell more work than try to compensate for the added expense by increasing the markup on your jobs for the remainder of the year.

Using a Decrease in Profits — The third possibility is to reduce the amount of profit you want to make. Those numbers would look like this:

1. Projected sales volume for year is $268,000.

2. Projected overhead expense for year is $74,220.

3. New overhead expense for year:
    $74,220 + $21,749 = $95,969

4. Net profit for year:
   10% × $268,000 = $26,800

5. New projected net profit for year:
   $26,800 − $21,749 = $5,051

6. New job costs:
   Total Sales − Overhead and Profit
   $268,000 − ($95,969 + $5,051)
   $268,000 − $101,020
   New job costs = $166,980

7. New markup:
   Sales ÷ Job Costs
   $268,000 ÷ $166,980
   New markup = 1.6049 or 1.61

By now, you know that I don't recommend this approach. While it may work mathematically, it isn't good business practice to work for only $5,051 net profit. Would you want to invest in a company with that kind of return? If you use this last method of assimilating your increased overhead, then you *are* investing your hard-earned money in that company!

You can use a combination of any of the three different calculations above, but it's far easier to use the increased sales approach. Did you notice that the increased sales approach is the only method that will also increase your profits?

Making a Second Budget Adjustment — Now suppose, in addition to the new truck, you decide that you need a new computer for your business. You also decide to hire a part-time bookkeeper to handle the bookwork drudgery created by all your new sales. And, while you're in a spending mood, you purchase a new pneumatic 16d-nail gun. All this adds up to an additional expense of $7,345 to be assimilated into your budget through the end of the year. Lets do the math to see what this does to your sales and markup.

1. Adjusted projected sales volume for year (after truck purchase) is $342,848.

2. Adjusted overhead expense for year is $94,969.

3. Overhead is what percent of total sales?
   $94,969 ÷ $342,848 = 27.7%

4. New projected overhead expense for year:
   $94,969 + $ 7,345 = $102,314

5. New projected sales volume necessary to cover new expense:
   New Overhead Expense ÷ Overhead Percent

$102,314 \div 27.7\%$
New projected sales volume = $369,365

6. New projected net profit for year:
   $10\% \times \$369,365 = \$36,937$

7. New projected job costs:
   Total Sales — Overhead and Profit
   $369,365 − ($102,314 + $36,937)
   $369,365 − $139,251
   New job costs = $230,114

8. New projected markup:
   Sales ÷ Job Costs
   $369,365 ÷ $230,114
   New markup = 1.60513 or 1.61

To assimilate the additional $7,345 of overhead expense, you must increase your total sales volume by an additional $26,517 ($369,365 − $342,848).

When you make a midyear correction, also increase the number of times that you review your numbers through the end of the year to be sure you're meeting your new budget projections. Don't let it get by you, regardless of the reasons. Going from July through November without checking your numbers is just asking for trouble. You may find at the end of November that your sales are down 10 to 15 percent or more from your projections. It's too late to sell the additional amount of work, build it and collect on it to make your numbers come out right.

Let me be perfectly clear about this: Doing the math and making sure all the numbers are where they should be isn't enough. *You must also make it all happen.* That's the difference between those who make it in this business, and those who don't. Less than one contractor in ten will take the time to both develop these numbers and then make them happen.

And the Formula Is . . .

Have you figured out the formula? If you have, it should look something like this:

New Total Overhead ÷ % of Existing Overhead = New Sales Volume

Now that was easy enough, wasn't it?

Making Adjustments to Your Markup

We should take a moment to review some questions and concerns that come up from time to time regarding markup in this business.

Chapter 3: Establishing the Correct Markup for Your Company

Lowering Your Markup

First, don't assume that your markup always goes up. There are many things that you can change in the running of a business that will cause your markup to go down. One is that you might be able to reduce your overhead expense. Or you could decide to reduce the amount of profit you're making. Maybe you're getting so efficient in the day-to-day running of your jobs that your job costs are coming down. So even when things are going well, you can make adjustments. That's one of the options you can keep open.

Adjusting Your Markup to Suit an Owner

Never adjust your markup just because you think your price is too high, or because you haven't sold a job out of the last seven or eight you've quoted. Trying to price your work based on what an owner says he or she is willing to pay for the job is not only foolish, it's suicidal.

Along that line, let me give you a little tip that will save you a ton of time and headaches. I've been on more than 3,600 sales calls over the years. From time to time, I run into a customer who will come right out and tell me on my first visit that . . . "this job should only cost X number of dollars, and that's all I'm going to pay." When you run into this type of person, you'll have no doubt about it. They usually know just enough to make them dangerous and probably very difficult to work with. When you find yourself dealing with someone like this, ask him or her how they arrived at that particular price for the job. In most cases, they won't be able to give you a logical explanation of just how they arrived at that "correct price." If you're satisfied that they mean well, but simply have the wrong information, you can try to educate them about the real costs of the job. But if they're simply trying to bully you into a price that they've arbitrarily chosen, don't deal with them. You can waste a lot of time trying to justify your numbers, but you'll seldom make the sale.

Adjusting Your Markup to Suit Other Contractors

Never price your work based on what you "think" other contractors are charging or what you think the "market will bear" or any other such reason. First, you have no idea what expenses other contractors have. You don't know what their rent is, what their insurance costs are, what they pay their staff, or even if they pay themselves. And you can guarantee that their markup is entirely different from yours.

Let's return to the role of remodeling contractor for the moment and assume that we've been in business for two years and things are moving along nicely. Our accountant tells us that we're making a profit (Hallelujah!). If we do a good job of record keeping for our business, we should only need to adjust our markup once or twice a year for the next year or two and after that only when a change in our overhead expenses dictates it.

78 Markup & Profit: A Contractor's Guide

Figure 3-5
Comparative job estimates

Job Estimate			
Company	Job Costs	Markup	Sales Price
A	$16,548	1.48	$24,491
B	$16,548	1.51	$24,987
C	$16,548	1.42	$23,498
D	$16,548	1.31	$21,678
E	$16,548	1.39	$23,002

Figure 3-5 is a quick comparison of five companies, including our hypothetical remodeling company. Each company has estimated the same job, arrived at the same job costs, and applied its own markup to the job.

If we're company A, and we decide to cut our markup to be competitive with company D, we'll be giving away $2,813. That's all of our potential profit on the job, plus $854 of the money we need to cover the overhead expenses assigned to that job. Bad plan!

We don't know whether the contractor for company D knows what he's doing when he comes up with a job estimate at that sales price. If he does know what he's doing, and his numbers are accurate, that's terrific for company D. But chances are, based on the high rate of business failures in the construction industry, many of our "competitors" don't know what they're doing. So why would we want to price our jobs based on someone else's selling price? If their numbers are wrong, and we blindly follow their lead, what does that make our numbers?

Negotiating Your Markup

Negotiating a price for the work you're going to do can mean different things to different contractors. I feel that you can negotiate *what* you're going to do, *when* you're going to do it, *how* you're going to do it, and *where* you're going to do it — but never *how much you're going to charge*. The minute you begin negotiating the price, you lose. Your selling price must be the price that you've calculated by assembling your actual job costs and then applying your markup. If you contract for less than your full markup, you lose.

Recouping a Giveaway

Let's assume again that you're a remodeling contractor. (I like to use a remodeling business for our examples because they have the highest rate of business failure. So by deduction, we can assume that many of them are guilty of these sins I keep harping on!) Your average job size for the last year has been $12,500, which means that your average job cost is probably about $8,334, or just over 66 percent. Your overhead expenses are $3,166 per job, or just slightly over 25 percent. You've been making an 8 percent profit using a 1.50 markup.

You've been out on a sales call to see Penelope and Winthorpe Authause about remodeling their kitchen. You do your sketches, your estimate, and prepare for the callback to present your proposal for the job.

You arrive at your appointment and make your proposal. Mr. Authause says he likes your proposal and would like to give you the job, but your price is $500 higher than the $15,000 quote they got from Kitchens Overnite. He says if you can match their price, the job is yours.

In a gush of enthusiasm, you drop your price $500 and write the contract. You rationalize it by telling yourself that you'll certainly be able to find someplace to make up that $500 you just gave away, and besides, they're such nice folks!

Now let's see what you've done to yourself. At $15,500, your overhead expenses are $3,875 and you'd get a net profit of $1,240 on this job. But not now; you gave away $500 right off the top. So how long does it take to recover the $500 you gave away?

If all your other quotes were exactly $12,500, your numbers would look like those shown in Figure 3-6 below.

The numbers show that you would have to mark up your next job, assuming that it was at least your average job size, 1.56 just to recover the money you gave away on the last job. Or, you could spread the recovery out over two jobs using a markup of 1.53. But is that the whole story?

Figure 3-6
Raising your markup to recover a loss

Raising Markup to Recover a $500 Loss				
Markup	Job Cost	Quote	Difference	Total Recovered
1.50	$8334	$12,501		
1.51	$8334	$12,584	$83	$83
1.52	$8334	$12,668	$84	$167
1.53	$8334	$12,751	$83	$250
1.54	$8334	$12,834	$83	$333
1.55	$8334	$12,918	$84	$417
1.56	$8334	$13,001	$83	$500

No, there are several other issues that come into play here, and you should consider them as well. First is the ethical issue: should you charge your next customer extra to make up for failing to correctly charge your last customer? Only you can decide if that's a valid issue.

Second, Murphy's law will likely kick in. If you cut your price on one job, the next seventeen jobs will all be smaller than your average job size. So now you're dealing with very small jobs and having to decide on each one how much you can add on to make up for the giveaway but still be able to get the job. There's a point at which your price will eliminate your company from consideration. There's only so much money that people will be willing to pay, even if you have a good reputation.

Third is the issue of correctly managing your sales. As a salesperson, you have a responsibility to yourself, your company and to your customer. You must clearly cover all issues with your customer before you prepare an estimate and make a price presentation to them. This will save you an immense amount of time and effort. Establish a budget for the job before you present your estimate and quote your price, then stick to it.

It's easy to spot a customer who'll try to negotiate a price after you've quoted it. And they can spot a salesperson with a weak commitment to their quote. If price is going to be an issue with a customer, it's best to find that out up front before you spend a lot of time doing the estimate. With careful, well-thought-out questions, you can identify your customer's needs and spot problem areas before they become a concern. Then you can make sure your presentation builds enough value into your product that you don't have to give away $500 to get the job. *If you correctly manage your sales, you won't have to try to manage your losses.*

A word of caution for those who think that pricing high and then negotiating their sales price down is a great way to sell. Customers are getting wise to that ploy. They ask, "Why didn't you give me your best price the first time?" or "Why did I have to say no to get your best price?" When you hear questions like that, it's time to change your presentation.

It takes "business maturity" to sell your jobs at the correct price. Few contractors stay in business long enough to reach this level of maturity, but those who do make a lot of money.

Problems to Watch Out For

Now let's take a look at some problem areas that may take you by surprise. Many contractors think they're doing everything right, and still find themselves out of money and about to go bankrupt. In reality, they may be doing any of a number of things wrong: using poor estimating procedures, incorrectly applying markup formulas, or simply using poor judgment. Let's take a closer look at each of these.

Chapter 3: Establishing the Correct Markup for Your Company

Using the Wrong Markup Formula

Here's an example of a markup formula that's used by approximately 20 to 25 percent of all contractors, and *they're losing money using it*. The difference between my formula and this one is subtle, but as we shall see, very costly over a period of a year.

Let's assume that we're still running our new remodeling business and work out our numbers like this:

1. Projected dollar volume built = $150,000
2. Projected overhead expenses at 25% = $37,500
3. Projected net profit at 8% = $12,000
4. Projected job costs (subtract lines 2 and 3 from line 1) = $100,500
5. Markup ($150,000 ÷ $100,500) = 1.49

Let's check our numbers:

6. ***Markup × Job Costs = Volume Built***
   1.49 × $100,500 = $149,745

Now let's look at the formula used by many contractors to come up with their job price. This is a real MONEY LOSER:

Sales Price = Job Costs + Overhead + Profit

Here's how they figure the math:

Sales price = $1,750 + (25% of $1,750) + (Profit)

(The first error is that the overhead is figured on 25 percent of the job costs rather than the total volume.)

Sales price = $1,750 + $437.50 + (Profit)
Sales price = $2,187.50 + (8% of $2,187.50)

(The second error is that the profit is figured on 8 percent of the job costs plus the overhead — which is already figured incorrectly — rather than on the total volume.)

Sales price = $2,187.50 + $175 = $2,362.50
Sales price = $2,362.50

The correct sales price using my formula is $2,608, so you can see that you would earn $245.50 less ($2,608 − $2,362.50) using this method. That really doesn't sound like that much, right? But let's look at the entire picture.

The formula in question is:

Sales Price = Job Costs + Overhead + Profit

Let's use the formula, plugging in annual figures, and see what the results are.

Volume Sales = Job Costs + Overhead + Profit
$100,500 + (25% of $100,500) + Profit
$100,500 + $25,125 + Profit
$125,625 + (8% of $125,625)
$125,625 + $10,050
Volume Sales = $135,635

Using my formula our total sales volume for the year is projected at $150,000. Using this other formula, we'll end up with an annual sales volume of $135,635. The difference is $14,325. That's money out of your pocket that's needed to pay overhead and other expenses. Your overhead doesn't go down if you make less money per sale. Your job costs don't go down. You simply make less but spend the same amount of money, and eventually, you'll go under. The difference in the formulas is subtle, but very important, and not to be ignored!

Using Cost Plus and/or Time and Materials Estimating

This section is going to be very short and very sweet. Actually, not so sweet if you're using or considering using this method. Like using erroneous formulas for markup, this approach may look good, but it will cost you money in the end. It is, for most of the people reading this book, a very bad approach to doing business in the construction world. There are a few exceptions, but for most contractors, it's a guaranteed money loser.

I've met many, many fine contractors in my years in this business. Most of the successful contractors were sharp, educated individuals. Each had a little different approach to business, but not one single successful contractor I know uses Cost Plus or Time and Material methods for estimating their jobs.

Let's look at one of the major problems with Cost Plus job estimating. How exactly does it work? Just for fun, you write down how you think it's figured. Let's see if this method holds up under a few questions.

Most definitions of Cost Plus that I've heard say that it's an agreement between you and your customer that says you'll charge him the cost of the job plus an additional markup for your overhead and profit. This agreement is supposed to be clear and determined before the job starts. Yeah, right!

I'll play the devil's advocate here for a moment. Let's say you've made this kind of agreement with a customer. OK, what is the exact agreement? Is it your job cost plus a fixed percentage of the job cost to cover overhead and profit? You've just seen what happens when you start adding percentages of your job costs to cover overhead and profit. You lose money.

Your customer has no idea in the world what your overhead and profit percentages are or how you arrive at them, and, to be perfectly honest, he could care less. But if you tell him that you're going to charge him job cost plus 42 percent to cover your overhead and profit, I guarantee he'll go ballistic on you. A majority of your customers think you operate on a 10 percent overhead and 10 percent profit margin. They'll tell you that anything over those numbers is way too high. Now you're catching onto the problem with this method, aren't you? How you arrive at the percentage for overhead and profit is not clearly defined — and you and your customer will probably never see eye to eye on that.

Here's a story that will give you another good reason to avoid Cost Plus job bids. Several years ago, a contractor I know did a job in the $250,000 range for a family. He noticed on several occasions that the lady kept a notepad handy at all times during the job. When he presented his final bill for the job, she presented him with a TIB list — a list of all the time that his crew had spent in the bathroom while working on the job! I kid you not! She had the names, dates, time in and time out for every bathroom visit over the five-month job period written down. Based on his estimated hours and crew rates (which included a 30 minute unpaid lunch and two 10-minute paid coffee breaks a day), she claimed he was overcharging her for the job. He ended up having to return almost $1,500 for his crew's "Time In Bathroom!" How do you argue that one?

Using Cost Plus or Time and Materials, you'll have discussions with the nice folks about why you're charging them for a 50-pound box of 16d nails for 15 lineal feet of wall framing. You'll have discussions about why they have to pay the entire amount of the 30-yard dump box when they saw your crew bring in two doors and a broken metal window from another job and throw it in the box. They'll question most if not all the labor bills that you present for the job. Someone will show up late or leave early, and they'll call you on it. If they think someone is trying to take advantage of them, they'll bird-dog your crew until the job is completed. Why do they have to pay for your crews to pick up and deliver materials? Shouldn't that be the supplier's responsibility, part of their service? While some of the questions may be legitimate, some, like questioning you about who pays for your crew's time in the bathroom, will drive you nuts. Is it worth it? Not to me!

I know I'll hear from people who say they operate successfully using Time and Materials or Cost Plus methods. There probably are a few contractors who've figured out a way to make these systems work. For those that make this claim, I have a couple of questions:

1. Did your company make a minimum of 8 percent net profit over the last year? (That's after you paid yourself for any and all work you did on jobs, and paid yourself a regular salary for running your business, and paid any family members a fair wage if they worked in the business with you.)

2. How many problems or disagreements about charges have you had with your customers on your last five jobs? Everyone has arguments with their customers from time to time, you say? Sorry, that's just not true. Not everyone in this business fights with his or her customers. Disagreements and possibly misunderstandings, yes; but arguments, no. In all my years, I've only had one disagreement with a customer, and it was due to his dishonesty (which was proven both by signed documents and witnesses to his statements).

If you can answer question # 1 yes and question # 2 none (you did make an 8 percent profit and you didn't have any arguments with your customers), then you're doing OK. If you answered question #1 no and question # 2 with some number, then maybe it's time to change.

Job Supervision and Markup

From time to time, especially if you've developed a reputation for good work, you'll have building owners or homeowners ask you supervise a building project for them. Their plan is to pay for all the labor, materials, subs and suppliers directly, while you ensure that the work gets done on time.

This is another subject that could take up a whole book. I'm just going to hit the high points. If you choose to take on this type of work, here are some suggestions to help you stay out of trouble. You need to know what to avoid, and most importantly, how to charge for this kind of work. Later in the book, when we get into writing contracts, I'll outline some important contract language that should be in any agreement that you write for job supervisory work.

First, why do you think they asked you to supervise the work? Plain and simple, they think they're going to save a buck. That should be your first warning sign. But will they really save money? Yes, I think in most cases they could save money. But the trade-off is a huge investment of their time.

If I were to take a job like this, part of the written agreement would be a designated time for at least one meeting each week. At this meeting we'd review everything that's been done to date, and discuss what's to be done in the upcoming week. There would be no loose ends left dangling. And the contract would also stipulate that I'd be paid for the time spent at those meetings as well as the time on the job! After all, the meetings are part of the job. The point is, these meetings will involve the customer's time as well as yours, and they must commit to that.

The customer may have worked in some area of construction, as a builder, as a developer, or in some other role that leads them to think they know more than they do. Be doubly careful of these folks. They know enough to make themselves very dangerous to the contractor who's working for them in a supervisory capacity.

I recently arbitrated a case involving a contractor who'd been hired to supervise the building of a $1.5 million residence for a real estate developer. The supervising contractor had agreed that if the cost of subcontracted work on certain parts of the job ran over the bids by more than a set amount, the contractor was to be held responsible for the overage on those bills. Everything was fine until the home was in the final stages, and the charges for additional work started coming in. The owner had been on the job site and had requested additional work from various subcontractors without mentioning anything to the supervising contractor.

During the arbitration hearing, I asked the owner why he had done this. He said there were two reasons. First, he wasn't always able to discuss the changes with the supervising contractor because he wasn't always at the job site when he (the owner) was there. And second, it was the "contractor's responsibility" to see to it that the subs did all the work he wanted for the quoted price. He had a few other reasons for taking this approach with the contractor, but those two were the main ones.

Even though I wouldn't normally do this as an arbitrator, I felt compelled to speak privately with the owner about his approach to dealing with this contractor. I told him, in no uncertain terms, that I believed his approach to this situation was dishonest. Since the owner had signed an agreement for binding arbitration, he agreed to pay all the additional costs for the work he'd requested. The job was finished under the supervision of the original contractor, and all ended well. Unfortunately, the owner had successfully used this tactic on other contractors in the past, bullying them into covering his excess costs. He thought that he could go on with this practice and it was perfectly OK, as long as he could get away with it.

I wish this story would be enough to scare contractors away from this type of work, but I know full well that it won't. So, let me give you a few brief guidelines for doing supervisory-type work. File this stuff in the back of your mind so that you can access it the next time you're tempted to run someone else's job for them.

Write a Good Contract

First, and most important, you need a written contract just as you would for any other job that you might take on. This contract must clearly state that:

1. The owner will only deal with you or your superintendent. They will make no separate work agreements with anyone else.

2. If the owner should become involved in the construction or make requests of any subcontractor, supplier or even one of your employees, they are solely responsible for any costs resulting from that involvement. In addition, they must pay you a fee in the amount

of at least 20 percent of the cost incurred by their involvement before you move on with the job. That penalty fee is what you might call an "attention getter!"

3. Payment for your supervisory work must be clearly spelled out, both the total amount, and the date and amount of any progress payments. You should also have a clause in your contract that specifically states that you can shut down the job if the payments to you aren't made on time. Again, we'll go over the specific language to use in contracts later on in the book.

4. The agreement should state that no changes are to be made to the contract unless they are *mutually agreeable*. This doesn't have to be longwinded, just plain and simple *mutually agreeable* will work just fine. Specify very clearly that any mutually agreeable changes to the job or to your agreement will involve a "change fee" to be paid by the owner based on some percentage of the total cost resulting from that change. You can make it 15 or 20 percent or whatever works for you, and include at least $45 per hour over your agreed-on price for any and all time that it takes to supervise that change. The change fee, along with any expense involved for architectural, engineering or design adjustments, or any research costs that it takes to implement the change, should be paid *at the next scheduled draw*. Never leave payment for changes to the end of the job.

Your Markup

Your markup for this type of work should be exactly the same as for any other job that you do. Estimate your time and the various costs involved in supervising the job, and then mark it up. Review the job with someone who's familiar with it to make sure you've accounted for everything. Then write your proposal. The proposal should say something like . . . "based on the plans and specifications as they are written this date, this is my price for doing this job, and here is how I am to be paid." You should get at least 20 percent up front before you start, and in no case should your final payment be more than 10 percent of the total for the job.

Some Don'ts for Supervisory Work

And finally, here are some things that you must not do.

First, don't negotiate a fee for supervisory work based on a percentage of the job, ever. If you do, you'll end up doing it for less than you should, guaranteed. *Remember, negotiate what, when, where, but never how much!*

Second, never provide the customer with a cost breakdown stating how you arrived at the price for your work. That's nobody's business but yours. When asked questions of that nature, I normally respond with . . . "That's proprietary information that I'm not at liberty to disclose!" I've even used that in a courtroom when asked by a judge how I charged for my work. He

Chapter 3: Establishing the Correct Markup for Your Company

looked at me kind of funny, said, "Oh, OK," then moved on to other matters. How you answer questions like that goes a long way in determining how much money you can make.

And third, never even consider renegotiating your contract once the job is started. This is something that you should be on the alert for, because I guarantee it will come up. You'll find that often, after you get the job started and it's going along very smoothly — especially if you're ahead of schedule, the owner may come to you and tell you that he wants to "renegotiate your agreement," or words to that effect. Because you make it look so easy, they don't think you've put enough time and effort into it. The fact that you know what you're doing and that you do it efficiently seems to escape these folks. Anyway, if they want to renegotiate your agreement, there are a number of ways to handle that situation. Here are a few:

- Tell the owner that, as stated in your contract, any changes will cost him 20 percent of the originally agreed-upon price. That usually stops any further attempts at lowering your fee. You might also remind him that you're getting paid for your ability to get the job done, not for the amount of time you spend at the job site. However, if he'd like to meet with you at $45 per hour to discuss changes in your contract, you'd be happy to go over the matter with him.

- Or, you could approach the situation like this: "You're right, we do need to renegotiate our agreement. I've had to spend more time on this job than I anticipated. I've been hesitant to bring it up because I believe in honoring my contracts. However, I've actually underestimated my costs by about 25 percent. I'm obligated to honor our original contract but if you want to change it, then we can account for the additional time and expense that I've incurred." Again, that usually puts an end to the conversation. However, if it doesn't, remind him that you will have to stop work while you renegotiate your contract (at $45 per hour for your time) and that may delay completion of the job and cost him additional startup and rescheduling fees as stated in your contract.

- And as your last argument, you might point out that there's a clause in your contract that states that any changes to the agreement must be mutually agreeable to all parties, and you don't agree to a reduction in your fees for supervising the work. That should be a conversation-stopper.

So now you can see how essential all these elements are to your agreement. They're all there for your protection — as well as the protection of the owner. Remember that all these points can be used against you if you should want to raise your fees. Make sure your estimate and your markup for the job are correct before you negotiate that contract. Then, if you feel that you can write a thorough contract and make money at job supervision, by all means, go for it. But be very sure that you'll make money. If you have any doubt, pass on it.

Chapter 4

Sell Your Services — at a Profit

"Nothing happens until somebody sells something . . . at a profit!" says Roger Dawson, author of *Power Negotiating for Sales People* (on audiocassette from Nightingale/Conant). Why am I talking about sales calls in a book about markup and profit? Think about it. If you don't get that first sale, it's not going to make a whole lot of difference what markup you planned. And the profit you were going to make won't matter much either. I'll spend this chapter going over some of the things you need to do before you have any jobs to figure markup on! A well-thought-out sales program, which starts with a good lead and takes you right through to the final contract with your customer, will allow you to charge more for your work and still get enough jobs to keep you busy. People are willing to pay more to do business with a company they trust, one that conducts business in a professional manner.

I can't overemphasize the importance of profitable sales in any business, particularly in construction. You can be the most skilled builder that ever lived, but if you don't realize that *Nothing happens until somebody sells something . . . at a profit!* you'll be the most skilled builder in the poorhouse. This phrase should be posted in the offices of everyone in your company, regardless of their position. Sales are just as important to the other employees in your company as they are to the sales staff. No sales, no money. No money, no job. It's that simple. So, how do you keep those profitable sales coming into your company? Let's take a look at some time-proven methods of bringing in business.

The Basics of Attracting Sales

Most construction companies get 90 to 95 percent of all their leads over the telephone. So, first and foremost, you have to think of a way to get that phone to ring. You'll have to do some form of advertising to let people know you're in business and tell them how to reach you.

Consider the idea: *advertising is what you do when no one is around.* Now that gets to the heart of the matter in a hurry, doesn't it? If the phone doesn't ring, nothing happens! So, when you haven't any customers, you have to work really hard to find some. You won't need to worry about how to calculate your markup if you don't have any work, so let's get busy and make that phone ring!

Business Cards: Don't Leave Home Without Them!

How many business cards do you have within arm's reach of where you are right now? What if someone said, "I'll give you a million dollars in cash if you can hand me a clean, fresh business card with your name on it in the next five seconds." Would you get the million? I'll bet practically all the contractors in this business wouldn't get it. Many contractors couldn't produce a business card in five minutes, let alone five seconds. Could you?

You should always have at least one business card with you, no matter where you are. Even if you're reading in bed right now, you could be using your business card as a bookmark. Most contractors, and I do mean *most*, are missing the best and lowest-cost advertising medium available by ignoring the importance of a business card.

How many people do you come in contact with each day that own a home or business — or both? Almost everyone you meet is a potential customer! Everyone lives somewhere. So why not let them know you're in the construction business, ready and willing to build them a new home, remodel their existing home, or build or fix up their commercial building? If you're serious about your business, then you have to let folks know about what you do. If you miss an opportunity to let someone know what you do and they could have used your services, you've just passed up business. And that's the same as taking the bread right off your kid's table.

I give out my business cards at the rate of about 2,000 every three months, sometimes more. I've ordered over 16,000 business cards so far this year, and it's only September. Ninety percent of them are gone, so it's time to reorder. I try never to let my supply drop below 1,000. If you were within 50 feet of me in the last three months, you probably have one of my cards! There's always a box of 200 clean, fresh business cards in my vehicle, another half-dozen both in my Day Planner and on my clipboard, and another half-dozen or so in my shirt pocket.

If I'm invited into any kind of meeting or conversation, I'll be ready to talk politics and business, religion and business, sports and business, business and business, or turkey and business — and always with my business card in hand. If you're in the construction business, get with it. Get your cards out and let folks know.

Make sure your cards are current, with your correct name, address and phone number on them. If anything changes, throw the cards in the trash and get new ones. Your business card and how it looks tells potential customers a lot about you and how you conduct your business. Clean, fresh, current business cards are mandatory. I know the phone company and the post office get their kicks by changing your area code or zip code right after you've just printed a whole new supply, and it's tempting to take your pen and cross out the old number and write in the new. But don't do it. It looks cheap and it'll make you look cheap. Those big office supply chains can do 1,000 cards for as little as $10. What are you saving?

You should also post your cards in places that you do business. Here are some prime spots for your cards:

- At your dry cleaners
- At the car wash
- At restaurants — always leave one with your tip when eating out
- Any meeting or social gathering
- A bulletin board at your place of worship or in their weekly flyer
- A bulletin board at the grocery store, feed store, discount store
- The lumberyard(s) or other material suppliers where you buy
- Your neighborhood business/home association newsletter
- Your children's school newspaper or yearbook

Include one in every bill you pay to a company in your area, and ask for a referral. The person on the receiving end may have a home or business that needs some building or remodeling work done.

Here's another area where many companies miss the boat. Every person in your company who's been with you more than 90 days should have their own business cards with their own name on them. *Everyone.* They see and know people that you don't — and with their own business cards, they'll be proud to show off where they work. The more people you can get your business card out to, the more opportunity there is for your phone to ring.

And if one of your employees brings in a lead that results in a sale, give that employee a financial reward. As soon as the down payment check for the job clears, award them a cash bonus, and make sure everyone knows about it. They've provided you with a very low-cost lead and deserve a reward for their company loyalty. The bonus should be a minimum of .5 to 2 percent of the sales price of the job — enough to keep all your employees interested in getting additional leads for your company.

There's an old saying: *The dullest pencil is better than the sharpest mind at helping you remember.* If someone has your business card, they don't have to remember who you are and what you do. In effect, you've written it down for them. That gives you the advantage over other contractors who won't be so easily recalled. Since you're the rare contractor who actually gives out your business card, chances are when someone needs work done, you'll be the first one to come to mind.

When doing seminars, I always ask everyone to pass their business cards up to me. In over eighteen years of seminars, I have never had more than 15 percent of the people in any audience pass their cards forward. Even when I say I'm having a drawing for a door prize, I get 85 percent of the names on scraps of paper and only a few on business cards. People just don't carry their cards with them. You can't pass out something that you don't have, or left at the office, at home, or in your truck.

Here's a quick exercise for you to do right now. Take a few minutes to make a list of at least six places where you could put a small stack of your business cards or where, in the next five working days, you could post your business card. Have you made your list? OK, tomorrow follow through and get those cards out there!

Every December I set goals for the coming year. This includes a complete set of goals for both of my businesses, as well as my personal life. I also have everyone who works with me do the same thing. Here are some of the goals we've established for Stone Construction Services:

- Leads per month
- Sales per month
- Total dollars sold each month
- Thank you cards sent out each month
- Contacts each month

That last one is probably the most important item for the long-term survivability of my construction company. My goal for this year is 762 contacts. At a ratio of six contacts for each lead, that's what it'll take to bring in the 125 new leads we want for the year. For me, a contact is any individual that I have talked with who's not a supplier, subcontractor or specialty contractor, service person, family member, or building or trade association person. However (and here's the key), I don't count the person as a contact unless they've been given one of our business cards. If they don't get a card, they won't remember our company or me in a few days. Ask yourself the name of everyone that you've met in the last seven days. Can you write them down? Probably not. We forget details. I do, you do, and they will, and all too soon.

Get yourself a business card holder. This is a three-ring binder with clear plastic business card inserts. Everyone who works for me has one, and so do I. Whenever you exchange business cards with someone, add them to your binder. If you want something done, you can go to your business card holder and look up that person. This business card deal works both ways. If you give out your cards and ask for one in return, you'll have something "in writing" to remember that person by. More important, you'll have a name and a phone number where you can reach them.

And, by taking their card, you've paid that person a nice compliment. You've let them know that meeting them has been important enough to you that you want to remember them. You form a small bond, and that can build into a good working relationship. The point is, you've made a contact and you've left a good impression. You never know if that other person will need your services or not. But if they do, you want them to have your card so they know how to reach you.

Job Signs

A job sign is another great form of advertising, and you should have one on every job that you do. Job signs specifically attract business from the neighborhood where you're currently working. And that's where you'd like to have more jobs. If there's a particular neighborhood that you'd like to work in and you don't have a job going on there, beg or rent space on somebody's front lawn, storefront window, or whatever, and put a sign up. If you work at it, you can find a way to get your signs up and that will give you an "in" for the area.

Make sure that your signs are attractive. Like your business cards, they say a lot about you and your company. Choose the design and colors for your signs carefully. The person who handles your advertising can tell you about using colors to attract the most attention for a particular area. You should be able to read the sign from a distance, so its composition is an important consideration as well. Moving signs from job to job can damage them. Protect your signs by making a cardboard sleeve to cover each one so when you move a sign, it stays clean and neat. Putting up a dirty, worn-out sign will send a negative message that won't bring you new business.

My company has "A-frame" signs that are 32 inches wide at the base and 48 inches high. They stand on 2 × 2 legs, bringing their total height up to about 4 feet 6 inches. The company name and other lettering are about 8 inches high. Signs should be big enough and the lettering large enough so that someone can read the company name and phone number from a minimum of 200 feet away. And follow the rule of thumb used for large billboard signs that you see when driving about town. They use no more than seven words. That way the message can sink in with one quick glance, taking only a second or two to read. The idea is to not distract the driver too long from the road, but rather to briefly gain their attention. They also use dark letters on a light background so that you can read them easily at night. If you put light colored letters on a dark background, the visibility goes way down. In fact, you're losing the use of your sign for a good portion of the day. You want the best visibility when people are out driving to and from work.

What if city or county codes don't allow you to put up job signs? Fine, put up a warning sign instead. With a little thought, you can devise a good warning sign. Right across the top put the word "Warning!" in large letters.

Below that say "Construction Zone" or "Construction Work in Progress" or "Remodeling Work in Progress." Follow that with "For information on this job contact ABC Remodeling" and make sure you include your phone number so someone can reach your company if they have any questions. Be sure your company name and phone number are in great big easily-readable letters.

If you're worried about how aggressively the community you're working in polices its rules about signs, you might make up a couple of signs with vinyl on cardboard to start. That way, if they continue to give you grief about your signs, and you don't want to fight the issue, you won't have lost a lot of money on signs you can't use. Just be sure your company name and phone number are clearly painted on your truck. They can't hassle you about *that*.

I know a contractor who decided to fight his city's signage regulations. He got an attorney and took the city to court. Although it took him over a year of legal battles, and several thousand dollars in legal fees, he finally got the city to agree to let him put up his signs on his jobs. He says he gets enough business off his signs that the fight was worth it. He even told me that he would go through the whole fight again if he had to.

Door Hangers

Placing door hangers around the neighborhood where you're working seems to work well for some contractors. The design of the hanger has a lot to do with the success of this undertaking. So put some thought into making it pleasant and appealing. Many contractors are reluctant to use this type of advertising. I don't know whether they're lazy or just afraid to go out onto the street and put them on doors. Maybe they're afraid the homeowners will come out and bite them.

If you do a good job on a remodel, the homeowners will want to show off your work to all their neighbors — as well as everyone else they know. If you've put door hangers around the neighborhood, that may get those neighbors over to look at your work. This kind of advertising is priceless. Good referrals are the best source of leads in the construction industry. Be sure to get at least three referrals from every job you do.

The Yellow Pages

Yellow Page advertising works for a lot of people. Some swear by it. You should have a listing with your company name in the phone book so people can find you. That's a given. You'll have to decide how much farther you want to go. Do some research about its effectiveness in your area. You'll be able to tell rather quickly how well it works, *if you'll do the research*. But I'm going to caution you about where you get your information. Talk to other contractors with companies that are comparable to yours.

Look for companies about the same size as yours, who have been in business about the same length of time. Find out how effective they think their phone book ads are. I wouldn't just take the word of someone from the phone company who comes to see you about advertising in their book. Guess what they get paid to do.

It's been my experience that the longer you're in business, the better your Yellow Page advertising will work for you. Name recognition has a lot to do with who people call when they let their fingers do the walking.

Direct Mail

Direct mail works for some, but this is a very tough advertising medium. You can only expect a response of from 0.2 to at most 2 percent of those who receive your mailing. Frankly, most of your effort will end up in someone's wastebasket. You have to really know your target audience to get them to open up that piece of mail. You'll probably need help from the people who do this kind of advertising for a living. If you're considering direct mail, go to at least two different direct mail seminars and see how the pros do it.

This type of advertising also takes a while to put together. You need to design the layout of the letter, the envelope and anything else you include, and then you have printing and folding time. It'll probably take three months minimum. If you're in a hurry, or on a tight budget, this isn't the place to start.

Direct mail can also be an expensive way to advertise, especially when you're starting out and you don't know exactly what's going to work. A direct mail piece may cost you $250 to $500 for your design. You'll need to have 2000 or more pieces printed to get your best printing rates. The printing costs for the total mailing piece, including envelope, cover letter, company brochure and return post card can easily run 50 to 75 cents each. The postage costs for a mailing of 1000 pieces will run you a minimum of $220, and if you load up the piece with five or six inserts, it can easily be double that amount. So your costs can run from $1000 to $2000 or more the first time you use direct mail. The costs will come down over time. If your design works, you can use it again and again. But more than likely you'll go through five or six revisions before you get a design that works well.

The post office can also get very fussy with your mailing pieces. You might plan to get them mailed at a certain cheap rate, then find out, after you've had them all printed and stuffed, that because something is too high, or too far left, or a quarter inch too wide, they have to go first class. When you have several hundred pieces, this can make quite a difference.

I've developed a mailing piece for Stone Construction Services, and after years of trial and error we now get a consistent 1.5 percent return on our mailings. A couple of times we've had a return of just over 2 percent.

We spend 90 percent of the company's advertising budget on direct mail. We mail 400 to 600 pieces every three to five weeks. That gives us a steady number of leads and eventually new work. We've spent a lot of time and money learning what works for us in our area. If you're going to use direct mail, expect to spend the money and invest the time to find out what will work for you, too.

Surviving Without Advertising

During my travels on the seminar circuit, I often run into individuals who pompously claim "We don't advertise. We do everything by referral!" Come on, let's get real here. If you want to do over $250,000 a year in business, you need a steady flow of new leads coming in. You have to go out and see a certain number of people each year to bring in a certain amount of dollars in work. That's the way of the world. If you don't advertise, in some form, you won't get the leads you need. That's just plain common sense. No leads, no sales. No sales, no jobs built, and that equals no money in the bank! The end of this equation is that your company is history.

Several years ago, I worked for another contractor as a straight commissioned salesman. Leads were coming in slow, and I'd been after him for a while to get an advertising program on line, which he'd promised to do when he hired me. For one reason or another, the advertising program just didn't seem to happen. Finally, I told him that he had to start advertising so that I'd have some leads to work on. The company needed work and we both had families to feed. I didn't leave him any room to wiggle out of this discussion, as he usually did when I brought this subject up. And finally the truth came out. "I don't want to advertise," he said. "It would be demeaning to my company!"

That was my last day on *that* job! The man didn't have a clue about the ultimate effect his prejudice would have on his company. He certainly didn't care about the effect that it had on the people who were counting on his company for their income. You can't hope to survive in this business with an attitude like that, and he didn't!

Your Telephone Is Ringing, Now What?

The phone rings and you have a "live one" on the other end of the line. They want to talk to someone about some remodeling, or a new home, or some other kind of work they've been thinking about. What do you do now?

How Is Your Phone Answered?

First, whoever answers the phone needs to clear their mind of anything but that call. Most contractors (and their staffs) aren't very good about this. If you doubt it, just get out the phone book and call three or four of your friendly local neighborhood contractors. It'll probably take you two or three minutes and several transfers or voice messages just to get to talk to a live body. More likely, you'll get to leave a message on a machine and maybe you'll get a call back. When you finally do get someone on the line who knows enough about the business to talk to you, it'll be another minute or two before you're sure that you have their attention and you can talk about your project. But before you actually get to what you want to say, they'll have to go find a piece of paper to write down the information that you want to give them. And somewhere in the background, you'll hear the dog barking and the kids screaming at each other and you may be put on hold while the person on the other end goes to stop a fight or close the door or turn down the TV or answer the other line. Sound familiar?

How well is your phone is answered? Quietly, quickly and by someone you know and trust, I hope. Do a check. Arrange for a neutral third party to call your company with a request for a quotation on remodeling work or whatever it is that your company does. Then get a candid and frank evaluation of how well the phone was answered and the information given by you or your staff. Good companies check out their customer response times and ask for customer feedback on a regular basis. *This is the most important customer contact point that your company has, and it needs to be right every time.* When you pick up that phone, your mind must be clear and you must be ready to talk to that customer about what's important to them. Put everything else out of your mind and listen to what they need.

Listening to the Customer

Listening is an acquired skill that few people take the time to learn. People in the construction industry seem to be particularly bad at this most important of all skills. I won't harp on this, but consider taking a class on improving your listening abilities. Many colleges and adult education programs offer classes on improving listening skills. They're well worth the investment of your time.

You've probably heard that you only have a very brief opportunity to make a favorable impression on another person. Most studies put the window at two to four minutes. Within that short time, you've got to make the other person comfortable enough to want to discuss their business with you, or at least willing to ask you some questions.

When you pick up the phone, be ready to listen and do some business. Have a pencil and a pad available to take down the information, or better yet, a contact lead sheet. Have your Day Planner, calendar or organizer handy so you can set up an appointment on the spot. Then *listen*! Put your opinions in your pocket and hear what they think they want. You can always guide folks away from bad ideas or designs when you meet with them, but on that first call, just listen.

Meeting the Customer

Let's assume that all has gone well to this point. You have an appointment to see a potential customer about some work. For our purposes, it doesn't make any difference what kind of job it is. There are some basic techniques in selling that apply in every situation. Make yourself a checklist to follow. That will help keep you focused, so you do everything you're supposed to do on your sales call.

Your Dress

How should you dress for a sales call? First, you need to be sure you're clean and fresh — if you have any smell at all, it should be a pleasant one. That involves taking a shower and maybe putting on some good smelling stuff, but not too much! Never go directly from a job where you've been working hard physically and meet with a customer. You can't offend people and be convincing at the same time! Make sure you plan time into your schedule so you can go home and get cleaned up before you meet with customers.

Be sure you dress in appropriate business attire. Gents, always wear nice slacks, a freshly ironed shirt with a tie, clean, polished shoes, and a sport coat if the weather is cool or cold. Women should also dress professionally. It's tough enough for a woman to sell construction services, so don't add to the problem by wearing casual clothes that undermine your credibility. If you dress professionally, you'll be treated as a professional. I know several new home contractors who require all their sales staff to wear suits whenever they meet with a client. That's not a bad idea for any contractor.

Men should have a haircut at least once a month, and it should be up off the ears and off the collar. A well-groomed, clean-cut look is always in style. There's a saying in professional sales circles: "As your hair line goes up, your income goes up!" And so it does.

I'm always amazed by contractors who trot out the tired old line about how their customers wouldn't buy from them if they wore a tie. How would they know if they've never tried it? Most of selling is attitude and perception (your attitude, and your customer's perception of that attitude). If you want to be viewed as a professional businessperson, you must dress and act like one.

Your Attitude

One of my former employers said this to me one day when I was going through a real funk: "Michael, sales is just three things, attitude, attitude and attitude. Your attitude stinks. Go home and fix it!" Try that on someone sometime. Believe me, it will get their attention — and quick!

The time to check your attitude is before you meet with your customer, not after you've blown the sale. Again, you have just a few minutes to sell yourself. If your attitude isn't right, then you've wasted a trip. When you go out to the customer's home or office, it's for just one thing: the sale. You should have nothing else on your mind. If you have a showroom or a model home that your customers will come to see, you should be mentally prepared to do business the minute they walk through the door. You're there to serve their needs, and those needs are your number one priority as long as you're with that customer.

Be On Time

Along with not returning phone calls, being late to appointments ranks right up at the top of the list of things that turn customers off. There are times when you may be unavoidably detained by a traffic jam, a severe problem on a job, or possibly a family emergency. If something like this comes up, call your customer as soon as you know you'll be late. There's no excuse for not making a call. Cellular phones are so inexpensive now that you'd be foolish to try to operate a construction business without one.

Let your customer know if you're going to be 5, 10 or maybe 15 minutes late. If you think you may be more than 15 minutes late, give the customer the option of rescheduling the appointment for another time. In all probability, they'll want to keep the appointment, but they'll appreciate the fact that you've given them the opportunity to change it. If you've done a good job of setting the lead, you'll have all the phone numbers you need to reach people and let them know that you're running late.

There's only one thing worse than being late and not calling. That's not showing up at all. A blown appointment is the very worst sales mistake a contractor can make. When you make an appointment, you're making an unstated promise that you'll be there, and on time. If you don't show up, you've not only broken your promise, but you've told them that they're not important. If you don't take them seriously enough to keep an appointment, how could they trust you to build a kitchen or new home? Why would they want to do business with someone like that?

I'd estimate that 20 percent of the work my company gets is simply the result of keeping all of our appointments, and arriving on time. The common excuse that "I just got too busy" is no excuse at all. People who use that excuse aren't too busy, they're too disorganized. And the prospect knows that. Why would they want a disorganized contractor to do their job?

I got a phone call yesterday from a woman who wants to have a small concrete patio built. It's not the biggest job in the world, but she has a need, and based on the subdivision she lives in, she'll probably pay in cash. I can always handle cash. She said ours was the eighth company that she'd called in the last two weeks. She had set up appointments with four different contractors and not one contractor had shown up. The other three companies that she'd called didn't even bother to return her calls.

I set up an appointment to meet with her at a definite date and time, and gave her three different phone numbers, including my personal mobile number, where she could reach me. Last but not least, I thanked her for her call and assured her that not all contractors are flakes. Good contractors return their phone calls and keep their appointments. Anyone care to place a wager on who'll be building her patio? And what company do you think she'll recommend to her friends when we're done?

You should always be available to anyone wishing to reach you, and you should be able to communicate with your potential customers at any moment. No one with a telephone is out of reach.

Be Prepared

What all of these suggestions amount to is being prepared when you meet with your customer. What does being prepared really mean? It means that when you step in front of a customer, regardless of where you are, you're ready to write up the order.

You're clean and dressed properly, your vehicle is clean and well maintained, and you arrive on time. You have all the samples with you that you'll need and your brochure books are neat, clean, organized and up to date. You know all of your company forms and can use them with ease, and you have plenty of contract forms ready and available. Your estimating system is in place, your numbers are current, your calculator is charged, and you're ready to write the order. You know exactly what kind of work your company can do, and you know when it's best to refer jobs out to specialty contractors. Now that you're prepared, what's next?

The Basic Steps of the Sale

There are four things you need to know before you can put a package together for a new customer. Here are the four questions that you need to ask:

1. What type and size of building (or remodeling) do they want to do?
2. When do they want to start?
3. Who will make the buying decisions?
4. How much do they want to spend?

When you meet with your customer for the first time, go through the questions carefully. Then *listen* to what the customer says, and *respond* thoughtfully to their concerns. You should end up with a signed contract in a high percentage of your sales calls.

What Do They Want?

Obviously there are some niceties that you have to go through before you get to the relevant question: What's the job? It may seem like a simple-enough question, but this is one of the toughest parts of a sales call to carry out successfully. Most of the time you'll get a lot of dreams and far-fetched ideas. But there's a way to get through the wishing and down to the real job. Ask your questions and then double-check the answers. Let me give you some examples.

Let's start with a residential remodeling job for John and Mary Jones. Their home is just too small for their growing family and they want to add some space. Most owners have some ideas about where an addition should be, what it should look like, and a rough idea for a floor plan. First, you need to check that they're all in agreement on these things. It's no fun to settle on a plan with Mr. Jones, then find out that Mrs. Jones wants something entirely different, and you're in the middle. Assuming that you're talking with Mr. and Mrs. Jones, your conversation might go something like this:

"As I understand it, you want to build an addition out the back of your home here, and essentially add 16 feet more to the existing family room. You also want to include a new laundry area and a bathroom, is that correct?"

Their answer to this question can initiate a whole set of new questions from you. Each question should make them think about various aspects of the addition, and steer them into making small commitments on each decision that they make.

"Why do you think this area is the best place for your addition? Have you considered any other areas of the home where you might want to build the addition?"

This might help you guide them away from a poor design choice and into an area where the addition would fit better, either structurally or aesthetically.

"Who will use this addition? Will it also be used as a guest bedroom?"

You might need to include design ideas that would provide privacy for guests — before the plans are set.

"Will you move the whole family in here to eat on Thanksgiving or Christmas or other holidays?"

Address the possible need for an extended space within a dining room, where they'll want to set up a large table.

"How many windows do you want? Why do you need that many windows? Have you considered using a skylight?"

"What type of door(s) do you want?"

"Do you want the exterior to match the existing look of the home?"

Well, you get the idea. Your questions can go on and on until you (and your clients) have a clear picture of exactly what their needs are and what they're willing to pay for. You must get them totally focused on the addition. Then at some point in the conversation, you need to switch the subject from *you* to *we*. It's a small, subtle change, but very important. This transition eases them into accepting you as their contractor and makes you part of their project.

When you've asked several questions about how "we" will do this or that, your questions might become a little more pointed.

"Do we need a bathtub or shower in the bathroom?"

"Will it be used by the whole family, or just by your guests?"

"Have you considered doing some of the work on this project yourself, like painting, wall papering or clean up to help keep the costs down?"

"Do we have any special needs or requirements for the electrical system?"

"Is the budget for this job the most importance factor, or are you more concerned about the overall look and design?"

With careful questions, you'll eventually get them focused in on what they want to do, but even more importantly, on *what they need to do*. I've found, after going on over 3,600 sales calls, that dreams and reality are often worlds apart. There's what they want, and then there's what they need (and can pay for). Who among us doesn't have champagne taste? Your job is to define their needs as quickly as you can.

I've spent time working with both Realtors and builders doing their own sales, and I think they both tend to make the same mistake. *They talk about money too soon.* Too often, one of the first questions they ask is "How much do you want to spend on your new home?" Give your customer some room. Let them get to know you, let them breathe free and talk to someone who's genuinely interested in their dreams. It might take you longer to get around to reality. But what's 45 minutes or an hour more if you win their confidence?

When you go looking at new cars, how do you feel if the salesperson starts asking you questions about money right away? You're not comfortable, right? Few people are. So don't do the same thing to your customers. Do something that few people in this business ever do; consider the customer's needs first!

If you use my sales process, doing a good job asking each of the first three questions, you'll find that the time to talk money (which is question number four) will come along naturally. Then you can move into that discussion with ease and comfort for your clients. With careful questioning, you can move the discussion in almost any direction you want it to go. With some practice, you even can get your customers to lead the way.

I always let my customers bring up the subject of money first, and sometimes twice, before we get into any deep discussions about it. Money is important, but you must talk about it on your terms and on your time schedule. You have things to find out about them first.

When Do They Want to Start?

The transition from "What do you want to do?" to "When do you want to do it?" should take place naturally. Here are some transitional questions that you might use:

"We've talked about what we need to do here and how you want the addition to look, but we haven't talked at all about your time schedule for this work. Do you have a particular date when you'd like the addition completed? Or do you have a start date in mind?"

Depending on their answer, you can go several different ways. Your purpose is to get them pinned down to a start date or a completion date. If they give you a date, double-check it with them: "Why is that date important?" Then explain the process you need to go through to get to the actual start date of the construction. Finally, give them an estimate of the time involved in the construction.

You need to be sure that the entire schedule works for them, from the start date to your completion date. Check with both the husband and the wife. Make sure everyone is in agreement on the time schedule. Don't get lazy and assume that because he says the schedule is OK, she's going to agree.

If a potential customer is reluctant to commit to a date, and won't even give you a general period or month, then you know that they're still shopping. The more vague they are, the more aware you should be that they haven't bought you or your company yet. They won't make a commitment until you have their confidence. Your sterling personality, charm, dress, good looks and great humor aside, you still must be their contractor of choice before they make a commitment.

At this point, depending on the job and the people that you're talking with, you've probably invested 30 minutes to $1^1/_2$ hours in this first meeting, perhaps even more. You should have the first two steps almost completed. Remember, the purpose of the first call is to find out the answers to all four questions that we listed at the beginning of this section.

Who Will Make the Buying Decisions?

Before you go any further on this sales call, you need to evaluate whether these people qualify to buy from you — not if you qualify to sell to them. That's one reason this four-step process is so important. Part of business maturity is the ability to know when you should stand up and walk away from a particular customer. You need to recognize when something isn't going well, and be able to tell that this job won't ever be right for you and your company. When you know when to walk away, you'll have reached a plateau few contractors ever arrive at.

This is sometimes a very tough call to make. You must be aware that you can't be all things to all people. Some folks simply have different agendas than you. Don't get emotional about it; just get up and walk away. The best advice I can give you on this subject is to ask yourself, "Is this a good customer for me?" or "Can we work with these folks?" Then listen very carefully to that little voice in your head. If it says "No," then give it up. That little voice, given a chance, will keep you out of more trouble than you can imagine. *Always trust your instincts.* I've learned to trust that little voice implicitly. I never doubt it, and I never regret those decisions that I've made based on my intuition.

By now, you've decided whether or not you can work with these customers. If you're continuing with this first appointment, you've decided you can. Now you've got to determine who'll make the buying decision. This not only includes who will make the final decision, but also everyone who'll be involved in that decision. This can be a sensitive area, so you must do some very careful questioning. You need to know who's going to make the buying decision, the couple together or a single individual. Don't focus on only one person when you deal with a couple until you know for sure who makes the decisions. That'll come out in good time. But until it does, include both in the discussion. Until you know for sure, assume that they'll both be making the final decision (although that's seldom the case).

Sometimes when you're dealing with a couple, one or the other dominates the conversation. Generally, it's the man who seems to ask and answer all the questions. When this happens, you must keep the woman involved in the process. Here's one way to handle that situation.

Suppose you're talking with a couple about a new kitchen remodel. You've asked a question that you'd like both to answer, such as "Where do you think the refrigerator should be located?" He jumps right in and says that it should go next to the family room door so he can get to the snacks quickly when he's watching Monday night football. You smile, turn to his wife, and say "Don't you think you might want it a little closer to the food preparation area, Mary? Where do you do most of your work in the kitchen?" You must get her response as well.

This is where you need to be strong. If she doesn't speak up right away or he answers for her, you must work around him without being offensive. That's the only way to get a design that's functional for everyone involved. Ask the question again, and direct it to her. The first time you do this, you'll probably catch him by surprise, and get a rather quiet answer from her. But she'll begin to trust that she's dealing with someone who thinks her ideas count.

The next time you ask her a question, repeat the process if he jumps in with an answer. It may take two or three times, but if you politely persist, he'll get the message and let his wife answer the questions you direct at her. Keep in mind that although she may not say much while you're there, if you ignore her and make plans without including her, she'll have a whole lot of input when you're gone. Ignored is exactly what you'll get, too.

Now suppose that the people you're talking to won't be the only ones involved in the decision-making. Maybe they're adding a mother-in-law unit and mom's going to pay. You need to be aware of this, even though they may not volunteer this information. But with careful questioning, you'll suspect that the project will involve another party. That's the time to ask if that third person will be involved in the decision-making process. If the subject never comes up during the course of your conversation, then you need to probe for the information, gently but firmly. Remember, no one likes to admit they can't pay for a new home or a remodeling project by themselves. There may be tender egos involved, so you must approach these questions with care.

You can spend a lot of time wondering, or save time by simply asking: "John, Mary, is there anyone else who'll play a role in your decision to do this addition?" If the answer is yes, try to find out whether that person should be at the meeting, or if you'll need to review all the numbers with that person before a final decision can be made. Do this in a matter-of-fact, but firm, manner. If the third party lives out of town, they probably won't be at your next appointment, either. If that's the case, you need to make sure that you have that third party 100 percent in your camp before you give any final numbers or make any commitments to the job.

"You need to recognize when something isn't going well, and be able to tell that this job won't ever be right for you and your company. When you know when to walk away, you'll have reached a plateau few contractors ever arrive at."

If the third party decision-maker lives locally, stress the importance of their being at your next meeting so you can explain the final details and review the sales price. If they're not there, your chances of selling this job are slim to none. I've been through that scenario about 30 times over the years. I'm still waiting for the first sale from any one of the calls where I quoted a price without all parties being present.

Believe me, if that third party doesn't have all the information before they hear the cost, they'll think they're paying way too much for the work. You can't depend on someone else to relay all of your information to that third person. Most people can only remember about 20 percent of what they've heard after 24 hours. So the next day when they call Mom to tell her about the job, she's only going to hear about 20 percent of what you covered in your presentation, but 100 percent of the cost. That won't be too impressive. Get all parties together, if at all possible, for the final decision-making process.

How Much Do They Want to Spend?

That brings us to the final question: What's their budget for the job? Too often, I hear remarks like, "I can't ask folks their budget for the job. They'll never tell me!" Or, "You can't ask the customer for their budget! If they give you a number, they'll think that you raised your price to match that number!"

Now let's talk reality here. If that's what you expect, that's the response you'll get from your customers. Guaranteed! But you can't deal effectively with a sales prospect without that information. You must establish a budget early on, or you'll invariably hear the famous old line, "Your price is too high!" Have you heard that before? Or how about the other universal favorite, "We want to think about it!"

So how do you go about asking your customers for their budget on this job? Again, you use some basic questioning techniques. Keep in mind that you're after the budget for a couple of reasons. The primary reason is that you want to be on the same page as your customer when you talk about the real cost of doing the job. Otherwise, you're going to waste a lot of time. Another reason is to help you guide your customers in their design choices. They can't be looking at a high-end job if they only have the funds for a moderately-priced job. If they're looking at natural marble countertops for their kitchen, they'd better be ready and able to spend the money to get that marble!

So how do you get the budget established? You can use either a one- or a two-step approach. If the first attempt doesn't work, then you need to come at them from a little different angle. But do whatever it takes to get the information. If you accept that premise, then I can save you hundreds of hours of work and frustration. Here's how you do it.

First, and most important, you must *lead up to* talking about money as you go through the sales process. I normally start during my first phone call. At some point in the conversation I'll set a time and date to come out and see them. Then, I'll add a comment or two like this: "Mrs. Jones, let me assure you that I'll be there on time. If I'm going to be more than a few minutes late, for whatever reason, I'll call you. When I arrive, we'll review the work you want done, and if we can meet your design requirements, we can talk about the materials you might like to use, the time schedule and your budget for the job. I'll come prepared to give you all the information you need to make a good buying decision, and I'll do my best to answer any questions you might have." Notice I didn't ask them about how much money they plan to spend. I just happen to let the phrase "budget for the job" drop into the list of things to consider.

Those few statements may seem simple enough, but they fill the customer's mind with ideas that they'll have to give some thought to. Now they know that they'll have to have some answers ready when I ask my questions. I've already mentioned that I'm going to talk budget, and just as importantly, I've let them know that I'm going to ask for a decision to buy.

Reword this to fit your own speaking style, but be sure you tell them up front that you're going to talk money when you get there, and that you're going to ask for a buying decision. After all, that's why you're in this business. You want to make money. Why hide it? Your up-front attitude will separate you from the host of order-takers in this business, and help to establish you as a professional.

When you've told your potential lead what they can expect when you come out, they may say something like this: "We've just started looking and all we really want are prices so we have something to think about." Or they may say, "We're just getting ideas now. We don't really want to do this until next spring (summer, after school starts, etc.)." If you hear this, you have a decision to make. It tells you a lot about the potential customer and where they are in the buying process.

"First, and most important, you must lead up to talking about money as you go through the sales process."

First, keep in mind that they need something or they wouldn't have called. Second, you don't have their trust yet, even though a friend or relative may have referred them to you. Third, you need to redirect their focus. It takes careful listening to really understand where they're at and where they really want to be.

When someone tells me they're just getting ideas now, I know that the minute I meet with them I need to begin planting the seeds of an agreement in their minds. I want them to know they can't expect me to give them design ideas for free. If they continue to talk about doing the job later, I always advise them that it could cost them more if they put off the job too long. Products go up in price, as may the cost of getting the money to do their job. Listen carefully so you'll know which way to lead the conversation during that first appointment.

At some point during the first hour of the meeting, I casually mention the subject of money, just to get them used to the idea of talking about money. If they ask me how much something costs, I say something like "I'm not too sure what that (usually in reference to some particular item) would cost at this point. Why don't we discuss all the work that you would like to do first, then we can review the financial aspects of your project. I need to get a more detailed picture of what we're talking about. Is that fair enough?" Now you've told them again that you're going to talk money, and you've asked their permission to do so — but not until you have all the information you need. Asking "Is that fair enough?" helps to relax some tension that may build up around the discussion of costs. It's a handy phrase, and used sparingly, can help relieve moments of pressure.

Speaking of pressure, what about the issue of high-pressure sales? I know that's starting to creep into the minds of some readers. Well, I'm not talking about high-pressure selling. I'm talking about focusing on getting the information you need to find out if your potential customers qualify to buy from you. High-pressure selling is pressuring the customer to buy something they neither want nor need. Asking good pointed questions so you can find out what you need to know is another matter entirely.

If you have any doubt about this, just read along and try not to form any judgments until you see where we're going with this approach. Not once in all the years I've been selling have I ever been accused of using high-pressure sales techniques. If I ever suspect that one of my questions might be the slightest bit offensive to a potential customer, I'll ask permission to ask the question. For example, if I want to ask about a customer's budget, I might say "Would you mind if I ask you folks a question here about the financing for this job?" Usually they'll answer "No, go ahead." Then I ask, "Do you already have your financing arranged for this work, or is that something you'll need help with?"

If you get evasive answers to these questions, then you know one of two things. You either don't have their confidence yet, or they may not be prepared to commit to anything but a low price. You won't know until you give the situation a little more time and ask a few more questions. There's no right or wrong way to handle this situation. You just have to keep asking questions until your instincts tell that you have a good potential customer worthy of your efforts.

"High-pressure selling is pressuring the customer to buy something they neither want nor need. Asking good pointed questions so you can find out what you need to know is another matter entirely."

Don't confuse hard selling with your need to answer our four basic questions. If you don't get definitive answers to these questions, you won't get a definitive answer to the most important question of all: "When would you like us to get started on this work?"

If you're asking enough questions and really listening to the answers, they'll eventually give you the opening you need to talk about their budget. They may say something like, "We're not sure if we can afford this!" Or "We don't know if we should stay here and remodel or buy a new house. It all depends on the final cost of this project." That's your opportunity to talk about their budget. Try this approach: "We've talked about a lot of things in the last few minutes, and you've just indicated, John, that you're not sure if you should stay here or move. Your decision will depend on the final cost of this job." Feed back what they've just said; that's a terrific lead-in. Then go on and say, "Let's take a minute to talk about money. I've found that lots of folks are uncomfortable talking about money, but you seem to really want this addition and you know you'll have to pay for it. Logically that seems to be what we need to talk about now. Tell me, what do you feel you can invest in your home and still have it make financial sense to stay here instead of moving? I'm not asking if you know what the addition will cost, just what you think your budget should be."

Notice that I've eliminated the "dodge" that probably 90 percent of your potential customers will throw at you when you ask them for their budget. Be sure to include the phrase: "I'm not asking if you know what it costs." If you don't use this phrase, you'll almost certainly get the tired old reply "We don't know what this job will cost." Then you'll have to come back with, "I don't know what this project will cost either, John, but my question was, what is your budget for the job? We need to establish that." Then keep quiet and listen to their answer.

The real problem with having to come back at them with this reply to get them to answer your budget question is that some folks may feel trapped, or even pressured. So it's important that you set the parameters for their answer with your first question.

Their response will tell you what you need to say next, and where the project will go. Their budget is normally 50 to 80 percent of what the actual numbers will end up being. Now you've got to act as an educator and explain the reality of their job and the costs that will go into it. I cover the subject thoroughly enough so that when I've finished they'll have a very close idea of what they're going to have to spend on the job.

If they still don't respond directly with a budget figure, then you need to give them some high and low parameters for their job, and let them select from these. If I'm talking about a remodeling job, I use an article that appears in *Remodeling* magazine every year (usually in the October or November issue) to help establish the budget. It's a survey of what various remodeling jobs cost in different communities across the country. It helps make homeowners aware of what a kitchen remodel or room addition really costs. I might even take it one step further and give them a low, medium and high figure so I can try to narrow down what they're willing to invest in their home.

If we're talking about a new home instead of an addition, I'll give them some high and low parameters, in terms of an "economy" home, a "middle-income" home or a "very well-appointed home, worthy of their status." That last one may be a bit flowery for some of your potential customers, but others will eat it up. I also give them a high and low square foot cost for homes in each of the three economic levels so they can figure out for themselves what their investment will be. I've found over the years that if I let *them* calculate the square footage price of their new home, the numbers they arrive at bring them to reality much more quickly.

"No matter what kind of contracting you're doing, use the same sales techniques. Get the owners to set a realistic budget, get them to make their selections, and you'll eliminate most of the potential sales objections when you ask for the order."

Here's another example of what I say, depending on how I read the people I'm talking to. "John, Mary, we can talk roughly about three general types of homes in terms of your investment. A very nice home, top of the line, with hardwood trim and floors, one that would be equal to any home in the area, will run you from $125 to $135 a square foot." (Of course, you'll insert the appropriate numbers for your company and your area.)

"Then there's what we would call a middle-income home, that will normally run from $95 to $120 a square foot. The exterior will look the same, but in this price range we're into a different grade of carpeting and vinyl on the floors, as well as different trim, and some of the other features will change, and some will just go away. But you'll still have a nice home.

"The economy approach to building a new home will cost you about $70 to $90 a square foot. Again, we're cutting back on some of the features to lower the cost, but providing you with a good, well-built home. You may be able to find contractors who are building starter homes at $55 to $60 a square foot, but from what you've told me about your needs, I think you folks will want a little more than the very basic house that you get for that price.

"You can do some of the work yourself, like cleaning and painting, and cut down a little on the cost. But I feel that if you're going to invest that much in a home, why not let the people do the work who make a living at it? They can do it fast and get it done right, while you do what you do best. The time you'd spend trying to save money by working on your house would be better spent earning extra money to get this home built the way you want it.

"So, with the information you have now, you can calculate the approximate price of the home you want to have built. Before we go on, however, you need to establish the budget for your home so that I can guide you through the selection process of the various items that will require your decision."

This is the point where many sales people make their critical mistake. They keep talking and don't get the folks to commit to a budget. Don't make that mistake. When you've asked them to make a decision and set the budget for the job, *stop talking*. Make them tell you what you need to know. If you let them slide on that decision, they won't make any other important decisions along the way. Then when you ask for the order, they'll slide around that as well. You'll have spent a lot of your time educating them to buy from someone else who has the sense to get them to make the important buying decisions as they come to them.

When they give you a number for their budget, double-check with all parties to be sure they agree 100 percent on the amount they're willing to invest in the home. If John gives you a maximum of $180,000 that they're willing to invest, repeat that price to Mary. "Mary, I want to be sure that we're all in agreement here about the budget for your new home. John has indicated that an investment of $180,000 is the figure that he feels is right. Do you agree with that figure?" Maybe you'd word it a little differently, *but the point is to double-check that number*. This is a must. If you do that, when you come back to them with your final numbers (assuming you've carefully guided them through the decisions that keep the final budget within their parameters), they won't be taken by surprise — and you won't hear "Your price is too high!" After all, they've already agreed to invest that amount of money in their new home.

Right now, you also need to pin down all the decisions about the job so you can give them your final quote for their job. This is where a good selection list is important. As soon as you know they're serious about the project, get them thinking about their selections. If you wait until you have your costs compiled and then try to go through the selection process, it doubles the time it takes to arrive at a workable estimate for the job. Sure, you can use material allowances in your contract to cover yourself, but isn't it better to get an exact quote for the work to be done? Also, I've found that if your clients won't take the time to make their selections, chances are, when you ask them to make the final decision to sign the contract, they won't do that either.

No matter what kind of contracting you're doing, use the same sales techniques. Get the owners to set a realistic budget, get them to make their selections, and you'll eliminate most of the potential sales objections when you ask for the order.

On simple jobs that you can estimate easily, you may ask for the order on your first visit. More complicated jobs will usually require at least a second visit before you present your total cost for the project. Here's another must: If you set the budget at one appointment, and then set a callback, *you must recheck that budget during the first few minutes of your next meeting*. "John, Mary, just to be sure we were all clear on the figures that we discussed at our last meeting, is the $180,000 budget for the job still valid? Is that still the amount of money you feel you can invest in your new home?" Again, clam up and listen to the answer. If anything has changed, it's better to find it out before you quote them a price.

Chapter 4: Sell Your Services — at a Profit

If they've met with another company that gave them a low-ball price for "the same job," they may have mentally adjusted their budget down. It's important to stay on top of the budget and continue their commitment to it, or you'll start hearing all kinds of excuses when you quote the final price. At that point it becomes very difficult to ferret out the real truth about why they're not ready to buy from you.

If their budget has changed, then you've got to go back to step one and start the sales process all over again. Forget the other company. Focus on reselling yourself and what you can do for them. Your professionalism will eliminate the "low ballers" almost every time.

Dealing with an Architect

Let me touch on a couple of items that are closely related to our topic. What do you do if an architect calls you to "bid" on a set of residential or commercial plans? It's the same story. You still must get a budget set for the job, *and it must come from the owner*. Architects are notorious for underestimating the costs of their designs. They'll deny this 'til the cows come home, but in all my years in this business I've only met two architects who could estimate a job with any degree of accuracy.

I recently bid on a two-story addition that illustrates this point. The architect who drew the plans assured the homeowners that the job could be done for less than $50,000. His "best guess," according to the owner, was $46,000 to $48,000. They got a job quote from their neighbor, a new-home builder, for $61,000. Later, he came back and wanted to raise his price. He said the job was too hard for his crews to do at the price he had quoted. They decided to look elsewhere. They got a quote from a "good remodeling company" for $85,000 and a quote from me for just over $121,000. I ended up looking like a real pirate on that bid, at least in the minds of the owners. But that was OK. I wished the owners well and moved on. I'd rather deal with people who are realistic in their price expectations. Someone's going to take a loss on this project, and that someone isn't going to be me.

So, when an architect calls you in to pick up a set of plans for a bid on a job, always ask about the budget. If they give you a number, be darned sure you ask if the number came from the building owner. Even if they say it does, I still take it one more step. I tell the architect that before I'll estimate any job and give a firm price quotation, I must meet with the owner of the project. I say that I need the meeting to discuss details of the work, the job expectations, and to be sure that I can work with the owner (and the architect, but I don't say that). I ask the architect to call me as soon as we have a definite appointment time. If the architect doesn't set up this meeting, or if the owner won't give us the time for such a meeting, I return the plans with a nice "thank you" and pass on the job.

Markup & Profit: A Contractor's Guide

Chapter 4: Sell Your Services — at a Profit

If they won't set a prebid conference, it's a good indication of a couple of things. First, and this is the case most often, they're simply looking to find the contractor that will give them the lowest possible price. Even when they find one who'll cut their price way down, that poor soul is still probably 25 to 30 percent higher than the architect's highest expected quotes for the project. Avoid those jobs.

Second, architects often want to be in full control of the project. That means they want everything to go through them. They don't want anyone else talking with the owner. Needless to say, I choose not to work with those people. If you find yourself in that situation, you must decide whether you want to be subservient to the master, or go find another job to do.

If you insist on a meeting with the owner and architect, make sure to use it to double-check the budget *with the owner*. That's not a time to be shy. Come right out and ask for the owner's budget. In many cases, I've discovered that the owner and architect hadn't even talked about a budget for the job. If they have, the number is probably several versions old and changes made since the beginning of the talks may have doubled the original price. That's why you can't be timid. *Don't leave that meeting without a firm budget set for that job!*

But here's a word of caution: Be careful when you talk to the owner about the budget. The architect, more often than not, has given them an unrealistically low number. If you make the architect look dumb in front of the owner, you'll alienate the architect. What will that accomplish? Think carefully about what you say before you say it. Give the architect all the room in the world to look good concerning their cost estimates for the job.

"So, when an architect calls you in to pick up a set of plans for a bid on a job, always ask about the budget. If they give you a number, be darned sure you ask if the number came from the building owner."

You may not get that job from the owner, and even if you do, you may not get any referrals from them. On the other hand, if you can establish a great working relationship with the architect, you should get plenty of referrals from him or her. Put your ego in your pocket and help the architect out. Almost all the architects that I've met are fairly bright folks. They're just not in the estimating business; they're in the design business. So, give them good numbers they can work with, and do it gracefully in front of their customers — especially if you detect that they know their figures are probably too low.

If the owner and architect are reluctant to set a budget, give them the opportunity to set another meeting so they can get their numbers together. You set the date and time for the next meeting, don't leave it up to them.

Markup & Profit: A Contractor's Guide **113**

This helps strengthen your position. If you give them three suggested dates and times for meeting and they still won't set a date, it's time to excuse yourself. Say something like this: "Ladies and gentlemen, we need to set a time to establish the budget for your job. You obviously haven't arrived at that point yet. You have my business card. When you're ready to sit down and set a firm budget for this job, give me a call. We would love to work with you on this project, but we can't if we don't have a budget to work from. We need to be able to advise you when a design choice or material selection will exceed your budget. We need parameters to work from if we're going to do the job right." Then get up and head for the door.

Now, if the owner and or architect does call you back, double-check to be sure they're really ready to talk budget. If not, don't set another appointment with them. Why waste your time? If they just have some more questions, then let them either pay you for your time or go get their education from some other sucker who doesn't mind giving his time and expertise away. Spend your time making money, not educating potential clients so they can go out and buy from some other guy who's 30 percent cheaper than you. This may seem a bit hard-nosed, but what else do you have besides your time and knowledge? You'll only make money when you stop giving them away.

Changes

When you've set a budget for the job with the owner and the architect and have everything down in writing that the budget covers, you're set. Then you can adjust your contract price every time they make a change. And trust me, they *will* make changes. What's more, every time they make a change on the job, they'll want you to throw the change in for free. "Oh, this is just a small change. We don't need to do an additional work order for it!" Right!

Write up a change order, put the new adjusted price on it, give it to the party responsible for paying the bills for the job, get their signature, and then do the change. If they don't sign the change work order, don't do the work. You don't have any obligation to change your original contract if they don't sign the order. If they delay signing the change work order, and the delay means you'll have to do additional work to put the change into effect later, write a new change order that reflects the additional charges for the additional work. If they won't sign the change work order this time, then stick with the original contract.

Follow the proper steps of the sale, do your homework, make the other parties do their homework, and you'll do profitable work. Get lazy, and you'll pay dearly.

Chapter 4: Sell Your Services — at a Profit

Some Final Notes About Sales

I've included some thoughts about sales here that could save you some problems later on.

One Call Closes

Recently there's been a big flap around the country about companies that do "one call closes." It seems that there are some in our industry, especially in remodeling, who think this approach to selling is something less than professional. I don't agree.

You're going out to see the potential customer to get the order; that's your job. If you're not, then you're either a professional visitor, or you like to waste time! It's been my experience that salespeople who spend time worrying about how "professional" the other guy is seldom produce much themselves.

The reality of sales is that you do what you have to do to get the order. If you can get the order the first time you see a customer, so much the better. If it takes several calls to close the deal, fine. As long as you're conducting yourself and your company's affairs honestly and ethically, and your actions don't generate any complaints, then do your job. I've had periods when I've closed as high as 40 percent of the leads that I went on during the first call. Today, I almost always end up going back two or three times before we get the final contract signed. The bottom line is this: If you conduct yourself in an honest professional manner, the sales will come to you and you'll do well and have happy customers.

Promised Rewards

Occasionally you will find yourself dealing with individuals who'll try to coerce you into giving them a "low bid" or the "right bid" with hints of future rewards. We see this most often when dealing with homeowners on small jobs. They say they'll be doing some grandiose addition next year and we'll get that job too if the price on this one is right. We used to get this kind of implied reward when we were bidding insurance claims as well. Don't fall into that trap. Take each job and bid it on its own merits. If you fall for the line that you may get "the big job if you're price is right," you deserve what you get. In most cases it will be a job that you'll lose money on, a fight with the customer, and no future work with that company, architect or family again. Compile each estimate realistically, with your fair markup, and you'll do well.

Written Bids

You'll find that today many potential clients request a written bid. This request, or demand in some cases, for a "written quotation" takes some thought. Do you want to participate in this game? They want your bid in writing so they can give it to their real estate agent, brother, banker, insurance adjuster, lender, nephew, or anyone they can think of. If you want to get involved in that nonsense, then by all means do so. It's been my experience that there's normally some hidden reason why they want your estimate in writing. They have a claim against the insurance company, or they're going to sell the home and want the information for potential buyers, or they're trying to get some money out of their mother-in-law. Whatever the real reason is, the result will most likely not be business for you.

When asked for a written bid, I respond, "Why do you need a proposal from us in writing?" I listen to the answer, and then tell them that we'll be happy to present a written contract. If they're ready to sign a contract with us to do the job, we'll gladly put that together for them. If they say OK to my proposal for a contract, then I've made the sale. If they insist on just a bid in writing, I tell them that I charge $75 an hour for a written estimate, and it will take a minimum of four hours to prepare.

If you think that's too tough, let me ask you this. Do you think that you'll get even one out of ten jobs where they hit you up for a bid like that? The public seems to believe that we're in business to "give out estimates" without any commitment from the person wanting us to do all that preparation work. One of the tried and true steps for successful negotiating is: *If I do something for you, then you must do something for me. Otherwise, what I do has no worth.* So, if someone wants me to prepare a quotation for them, then they must either be ready to sign my quotation (my contract) or pay me for my time to prepare the quote.

I have a phrase I use from time to time that really gets the customer's attention. If they start asking me to do something for them, and don't offer to pay for it, here's what I say: "When you go to your job each morning, you get paid for the services you provide, right? I feel I should be paid for any work that I do as well. I'm happy to help folks with ideas and suggestions on their projects, and tell them what it will cost to do the work. But what you're asking me to do goes beyond that and falls under the category of my working for you. And for that I should get paid." This is just something for you to consider the next time someone asks you for a written bid or anything else that involves more work than you want to do for free.

Handling Objections

I'm often amused by articles I see in national magazines (written by well-meaning experts) about how to handle objections when selling construction services. While I do believe that the term *objection* can describe a particular reason that folks may offer for not signing your contract, I also believe that you can structure your presentation to eliminate most of these "objections" before they occur. We've already discussed how to eliminate the "your price is too high" objection from your presentations. So while some salespeople might call that an objection, I call it simple laziness. If you hear that objection after reading this book, you're not following through on the sales steps that we just discussed. You're not:

1. Getting the budget established and double-checking it with all parties to the final decision before going to the trouble of putting an estimate together

2. Double-checking all selections with each person as they are made

3. Getting a commitment for a final decision when you make your final presentation and quote your final price for the work to be done.

Taking these steps will also eliminate the "We want to think this over" response that bedevils most salespeople.

If you plan and prepare carefully, and pay close attention to the responses that you get from your potential customers, you can build a presentation that will eliminate most of the objections that you've been hearing. A very good rule of thumb in sales is this: *If you get the same objection more than three times in a row, you're doing something wrong or you're leaving something out of your presentation.* It takes careful attention on your part to pick this up, but if you invest the effort, you'll find your sales ratio and your profits going up.

The "New Angle"

Now let's talk about this "new angle" that you hear about that's working its way west from the East Coast. It's the practice of using a higher markup on the labor for your jobs and reducing the markup on other parts such as materials, subcontractors or rental costs. I'm not sure how "new" this approach is, or just who started it. The rationalization is that it "makes it easier to explain how we charge for our work!" Those who use this approach think their charges are more believable to their customers if they share this information.

If customers want me to start explaining how I charge for my work, I ask why they need to know. If they really want all the information about how I arrived at a final price quotation for their job, they should be willing to pay for it. If they don't want to pay, then they can go out and assemble those numbers themselves. I've spent well over 35 years learning my trade and I'm not in any hurry to give away my time or expertise. Does my attitude on this subject cost me sales? I'm sure it has, and in all probability, will again. That's a choice I've made. I don't want to get into arguments or discussions about whether my price for some item of work is too high either before or during work on my jobs. The approach that I've chosen to take eliminates that problem right from the start.

On the other hand, if you're using the new angle and it works for you, then by all means continue on your merry way. If it ain't broke, don't fix it! Personally, I didn't like that approach and when I tried it, it didn't change my sales ratio a bit. I think it's a waste of time. This book is about developing your markup from your own company numbers, and using it across the board. Remember K.I.S.? Fussing around with raising this or that markup on some phase of your job and reducing the markup on other parts of the job seems like a lot of energy spent just to get from point A to B.

On the other hand, some contractors are now using the approach of explaining every number to their potential customers. Why would that be necessary? My premise is that if the numbers on their job actually meant anything to customers, you probably wouldn't be necessary. Why would they need you if they could do everything themselves? Again, get paid for what you do.

Here's the bottom line. If you're using a particular method of sales and you're selling at a good ratio, then by all means stay with it. As salespeople, we must each contend with the reactions our personalities generate from our customers. Something that works well for you may not work at all for someone else. If you own a company, you have a right to expect your sales staff to produce a minimum number of sales per leads given out. But you need to give your staff lots of room to create their own sales approaches. Sharing prices with customers falls easily into that category.

Low-Balling Estimates

This approach to sales has been around forever, but it seems to be more common of late. It's the practice of low-balling estimates, particularly allowance amounts. It's used to get the customer to pick one company over the others based primarily on price. Then of course, after the job is awarded to the lowest bidder and the work is started, the owners find that to get the quality of materials they want installed, they'll have to pay more. The cost for upgrading items that weren't selected in advance drives the price way up — usually higher than the more realistic quotes that they turned down.

I've run into this practice on at least one-third of all the remodeling sales calls that I've been on, and well over two-thirds of all new home construction sales calls. The first time I heard of this practice I thought it was dishonest, and now, many years and several thousand sales calls later, I still think it's dishonest. Those who use this approach only sell their services once to any customer. You can't build a business on unethical practices. No one will do repeat business or give referrals to a company that put the screws to them like this.

Today's customers aren't stupid. They may be dazzled for a brief period of time by the lower price, but at some point along the way, the high price of all the "extras" or the items "not included" will wake them up.

Giving someone a $6-a-yard allowance for floor covering or $10-a-square allowance for roofing materials when you know that'll it will cost much more for decent materials is just plain dishonest. Your customers don't shop for these items every day, and probably don't have any idea whether or not $6 a yard is a reasonable price for floor covering. But you know it's going to be junk — if you can find anything at that price! Wallpaper is more expensive than that, and it doesn't get walked on! Why in the world would you risk your career over something like this? Suppose you give a customer a $1,200 appliance allowance. They shop around and find out that the refrigerator with the ice and water dispenser in the door that they want costs $1,450. And they still have to buy a stove or cooktop and oven, a dishwasher, a garbage disposal and a microwave oven with that allowance! They'll either think you're stupid and you don't know the cost of things, or that you're trying to cheat them. Either way, you've lost their respect and you're in for a fight. Why put yourself through all that?

I love it when I discover that another company has low-balled an item or estimate on a job that I'm involved with. I focus my presentation on this "dishonest approach" to business. I suggest the customers go shopping and check out the prices of the items they're getting allowances for. They figure out for themselves real quick who's being honest with them. Sometimes you just need to point them in the right direction.

Always let the customer decide if they can afford something or not. Tell them the measurements or amount that they need, and let them go out to look. They'll figure out for themselves what grade they can afford.

If you've ever been tempted to try this "low-ball" approach to sales, don't. If you're using it now, stop. It'll destroy your reputation and your business faster than almost anything you do. Always leave a customer or potential customer with a feeling of comfort and mutual respect so that you're always welcome in their home whether they buy from you or not.

Down Payments

There's a lot of bad information going out over the airways, on both television and radio, by well-meaning but misinformed consumer advocates who say that owners should never give a contractor a down payment for the work they propose to do. I just heard this very statement on my car radio today as I was driving home from a sales call. It was the third time I'd heard this same guy ranting and raving on his talk show about the "grave consequences" of homeowners making down payments to their contractors. His reasons, though well-intended, simply boiled down to the idea that contractors can't be trusted. So how do you deal with this stuff?

As a contractor, it's important that you remember that your company is not a mortgage company, bank or lending institution. You're not in the business of lending money, so why should anyone expect you to start and do a job out of your own pocket?

About 3 percent of the contractors cause 99 percent of all the problems you hear about on TV and radio and read about in the newspapers. So why should the other 97 percent of us have to pay for the way that a few flakes in our business operate? We shouldn't!

The reality is that in the construction business there are expenses that have to be met up front, long before we ever get to the job site. Anyone who knows anything about the basics of construction knows this. There are architecture or design fees, engineering fees, plan checks, permits, impact fees, utility fees, up-front money for sub or specialty contractors, and the list goes on.

How many people buy homes in America without an "earnest money" agreement — in other words, a down payment? How many cars are bought or ordered from car dealerships without a deposit? None! So now we come to either the second or third largest purchase most home or business owners will ever make, and someone says that no money should change hands up front? That's ridiculous! Why should we in the construction business put up money out of our own pockets to get work started on someone else's property? We shouldn't. There's another saying in our business, "It's the wise and prudent contractor who works off the other guy's money." Not only is it wise and prudent, but it's the only fair way to do business!

I've only had three people in the last 28 years tell me they wouldn't give me a down payment for the job that I was going to do for them. I politely but firmly gave them the reasons that I required a down payment. One of the three agreed a down payment was fair and gave me a check. The other two decided that they wouldn't give me a down payment, and I gave their business to our competition. Apparently, there are "contractors" out there who are dumb enough to get involved in this kind of nonsense.

Here's one last item that the consumer advocates use. As a recourse for not getting money up front, they tell you that: "If the customer doesn't pay you for all your costs, you can always file a lien on their property!" Yeah, sure, you can do that. But who wants to? Filing liens can cost you $250 or more, and it doesn't guarantee that you'll get your money, at least not without a lot of trouble. The customers know that they can sit on the money they owe you for a long time. And, short of starting foreclosure action on their property, there isn't a darned thing you can do to force them to pay you. Now that's the reality of the situation. The advocates say you can file a lawsuit, but how much more will that cost you? Until it's settled, it's time and money out of your pocket. And the courts often side with the "poor victim trying to stand up to the crooked contractor." My advice on this subject is short and sweet: If the customer won't give you a down payment, walk away from the job.

Chapter 5

Writing Contracts

We're going to start this chapter by dispelling some "old wives' tales" about contracts. Some contractors claim that they can only use a one-page contract because their customers won't read anything longer. Others say that long contracts scare customers off. What these contractors are really saying is that *they* don't like long contracts, so any contract longer than one or two pages is bad.

I've learned some valuable lessons from some pretty smart people since I started in this business. And one of them had to do with writing contracts. What I learned is this: Short contracts will cause you more grief than almost anything else in this business. It's my opinion that short contracts are nothing more than lazy contracts written by lazy people who don't want to take the time or initiative to learn the value of a well-written contract.

During my early years in the business I sold remodeling for Jerry Jones of J & J Construction in Portland, Oregon. Jerry was the best remodeling salesman I've ever met or worked with. When I started with his company, he spent about six months helping me to develop and improve my sales technique. He could sell anything. During one period I saw Jerry sell 13 leads in a row. There were room additions, kitchens, bathrooms, siding and storm windows, a house leveling and a couple of large miscellaneous jobs in that group of sales. Jerry didn't miss on a single one. It didn't make any difference what the lead was, when it was his turn to take a lead, he sold it. Jerry was a great teacher. He was impeccably honest, hard working, detail-oriented and above all, a student of this business.

One day during my training period, I turned in a contract for a room addition. The contract was about a page and a half long. Jerry looked at the contract for a minute, asked me to sit down and said quietly, but firmly, "To a large degree, customers judge the value of their jobs by the weight of the contract. Would you rewrite this for me? And this time, write a contract that you would want written for you if J & J were to come into your home and build a room addition for you. I want you to cover every aspect of that job so that when I'm done reading it, the only question I will have is, 'Do you want cash for the down payment, or is a check OK?'"

I rewrote the contract. It turned out to be nine pages. I took it back to the customer for another signature and found that the customer was extremely impressed by my thoroughness. That was a very valuable lesson. Since that day, the shortest non-handyman type contract that I've written is eleven pages.

The Importance of a Detailed Contract

Customers normally respond to any idea or suggestion that you give them. About 85 percent of the decisions that customers make on their construction projects are influenced by their salesperson. If you lead them to believe that a long contract is bad, they'll believe it. On the other hand, if you spend a couple of minutes during the first meeting with your customers explaining the benefits of a well-researched and well-written contract, you'll eliminate any objection to a detailed contract before it becomes an issue. When you present the contract for review and signature, your customers will understand that they'll need a little time to go over the contract before they sign on the dotted line. They'll be expecting it.

And here's another important point. People buy when all their questions have been answered. There's no way in the world that you can write a contract for a kitchen remodel, let alone a new home or commercial building, and cover all aspects of the construction in one or two pages. Never use those preprinted contract forms (the ones where you fill in the blanks) that so many new home construction contractors use. Those forms are far too general and full of holes to be taken seriously by a professional construction company.

"You need to educate yourself and your sales staff on the correct wording to use in legal documents and then have the discipline to include that wording in every *contract."*

When you put a little "meat" into your contracts, your customers will perceive that as an indication of the value of your service. You could say that the heavier your contract, the more it justifies your markup.

What's even more important for you is that the more detail you have clearly outlined in your contract, the fewer arguments and disagreements you can get into. Plain and simple, if everything that you will or won't do is clearly written out in your contract and everything that's expected of the customer is clearly outlined as well, then you'll eliminate most misunderstandings. And if you should end up in court for any reason, a detailed contract will add weight to your case and increase the odds of a decision in your favor. Most judges make their ruling based on the contract. If the contract doesn't cover some specific item or issue, judges will tend to favor the home or building owner because they aren't the professionals and so shouldn't be expected to understand everything. The builder is expected to know what should be in a contract.

Remodeling magazine columnist F. J. Simon wrote that problems between contractors and customers almost always "arise from lack of clear contract language between the parties." Well said!

Putting the Contract Together

There's far more to writing contracts than just putting a bunch of words on a page. Be careful not to try to be a "shade-tree lawyer." This can be costly. Whenever there's the slightest doubt about any item in a contract, you should check with your company attorney before it goes to the customer.

You need to educate yourself and your sales staff on the correct wording to use in legal documents and then have the discipline to include that wording in *every* contract. You must be able to write detailed specifications for all the actual work to be done on a project. You can refer to the blueprints for the job up to a point, but you must create a clear word picture for your customer to see. The main reason for this is that there are often several versions of blueprints on the same job. I'm sure you've heard the famous line, "That's version five, we've been on version six for over three weeks now." Besides, most customers can't read blueprints; they need everything spelled out for them.

You should have a standard opening and closing for your document, plus several pages detailing the work to be done. Include copies of any city, county or state required documents, and of course, the federal "Right of Rescission" form. If you aren't familiar with the "Right of Rescission," or "Notice of Right of Cancellation" form as it is sometimes called, I've included one in Figure 5-1. Once you have all this together, add a nicely printed multi-colored cover to complete the package. Make three copies: one for the customer, one for the office and one for the field superintendent who'll build the job.

Our basic remodeling contract contains eight pages, minimum. The contract is expanded as the cost of the job and the amount of work increases as follows:

- Under $10,000   8 to 12 pages
- $10,001 to $20,000   12 to 16 pages
- $20,001 to $30,000   16 to 18 pages
- $30,001 to $50,000   18 to 24 pages
- $50,001 and up   24 pages plus an additional $1/8$ to $1/4$ page per $1,000

NOTICE OF RIGHT OF CANCELLATION

DESCRIPTION OF GOODS SOLD AND/OR SERVICES TO BE PERFORMED

NOTICE OF CANCELLATION   Date _____
(ENTER DATE OF TRANSACTION)

YOU MAY CANCEL THIS TRANSACTION, WITHOUT ANY PENALTY OR OBLIGATION, WITHIN THREE BUSINESS DAYS FROM THE ABOVE DATE.

IF YOU CANCEL, ANY PROPERTY TRADED IN, ANY PAYMENTS MADE BY YOU UNDER THE CONTRACT OR SALE, AND ANY NEGOTIABLE INSTRUMENT EXECUTED BY YOU WILL BE RETURNED WITHIN 10 BUSINESS DAYS FOLLOWING RECEIPT BY THE SELLER OF YOUR CANCELLATION NOTICE, AND ANY SECURITY INTEREST ARISING OUT OF THE TRANSACTION WILL BE CANCELLED.

IF YOU CANCEL, YOU MUST MAKE AVAILABLE TO THE SELLER AT YOUR RESIDENCE, IN SUBSTANTIALLY AS GOOD CONDITION AS WHEN RECEIVED, ANY GOODS DELIVERED TO YOU UNDER THIS CONTRACT OR SALE; OR YOU MAY IF YOU WISH, COMPLY WITH THE INSTRUCTIONS OF THE SELLER REGARDING THE RETURN SHIPMENT OF THE GOODS AT THE SELLER'S EXPENSE AND RISK.

IF YOU DO NOT AGREE TO RETURN THE GOODS TO THE SELLER OF IF THE SELLER DOES NOT PICK THEM UP WITHIN 20 DAYS OF THE DATE OF YOUR NOTICE OF CANCELLATION, YOU MAY RETAIN OR DISPOSE OF THE GOODS WITHOUT ANY FURTHER OBLIGATION.

TO CANCEL THIS TRANSACTION, MAIL OR DELIVER A SIGNED AND DATED COPY OF THIS CANCELLATION NOTICE OR ANY OTHER WRITTEN NOTICE, OR SEND A TELEGRAM, TO:

_____
(NAME OF SELLER)

AT _____
(ADDRESS OF SELLER'S PLACE OF BUSINESS)

NOT LATER THAN MIDNIGHT OF _____ .
(DATE)

I HEREBY CANCEL THIS TRANSACTION

_____   _____
(DATE)   (BUYER'S SIGNATURE)

IMPORTANT INFORMATION ABOUT YOUR RIGHT OF CANCELLATION

"YOU, THE BUYER, MAY CANCEL THIS TRANSACTION ANY TIME PRIOR TO MIDNIGHT OF THE THIRD BUSINESS DAY AFTER THE DATE OF THIS TRANSACTION. SEE THE ATTACHED NOTICE OF CANCELLATION FORM FOR AN EXPLANATION OF THIS RIGHT."

CUSTOMER ACKNOWLEDGES THE RECEIPT OF TWO COPIES OF THIS "NOTICE OF RIGHT OF CANCELLATION"

Signature of Customer _____ Date _____ , 19____
Signature of Customer _____ Date _____ , 19____
Address _____

CITY   COUNTY   STATE   ZIP CODE

CUSTOMER COPY

NARI

Figure 5-1
Sample Right of Rescission form

New home construction contracts should probably start at about 12 pages, and be adjusted up depending on the type of home you're going to build and where you're going to build it. Since I don't do new home construction, I hesitate to prescribe a particular number of pages based on sales price as I did for remodeling. There's a huge difference between remodeling and new home or commercial construction. Just make sure that your contract thoroughly covers all aspects of the work you are to do.

One final word for those who are still not convinced that a one- or two-page contract is dangerous: One of the most devastating things that can happen to anyone in sales is to go to all the work of meeting with the client, creating the design, making an estimate, and putting the contract together, only to have the customer cancel by means of the Right of Rescission. That can mess up even a good salesperson for weeks! When I first started out, I averaged one cancellation for every three contracts that I wrote. Then I had that contract meeting with Jerry Jones and the cancellations ended immediately. Since I started writing detailed contracts, I've only had one customer in the last 20 years cancel a contract using the Right of Rescission. I eliminate all their doubts and leave them feeling at ease.

Are you are experiencing problems selling jobs? Is your average lead-to-sales ratio higher than one in four? Are you getting cancellations on contracts you've written? If you answered yes to any of these, then I strongly suggest you take a hard look at the contracts you're writing. Make the effort to provide your customers with a good, solid, well-written proposal or contract. *Good contracts are the cornerstone of a professional salesperson and construction company.*

Your Attorney

Before we discuss the language of contracts, let's take a quick look at your relationship with the legal profession. You need to have a good attorney that you can call on for advice when you need it.

Attorneys see things from a different point of view than we do. They spend most of their days dealing with problems created by poor communication between their clients and others. So when you have a problem, chances are they've either dealt with the same situation before or they've studied a similar situation. They'll have a pretty good idea about how to get you back on track and on to the business of making money.

How do you find a good attorney? Get recommendations from people you trust, talk to a few attorneys and find someone you'll feel comfortable dealing with. In my opinion, there are two types of attorneys, "problem finders" and "problem solvers." If you talk to an attorney and sense that he or she is a problem finder, find another attorney. How can you tell which type they are? A problem solver will almost always look at the situation from the financial impact that it will have on your company. They listen to the basics of the problem, then ask how this problem relates to the company financially. If the situation will cost you money, they'll advise you to settle the matter as quickly as possible and move on with the least amount of expense. The problem finder will always have one more thing to check on, one more letter to write or one more phone call to make. The meter keeps on ticking, and guess who pays the bill?

One time I found myself doing battle with a company that refused to pay me a rather large sum of money. The situation continued for 29 months. Finally, my wife said, "Enough is enough. Your attorney is doing nothing but running up our expenses. He's no closer to settling this than the day he started. As of today, he's off the case. I've made an appointment for us with another attorney who was referred to me by a business associate." This was a bit tough for me as I considered my attorney a good friend. However, my wife had a good point, and we went to see the new attorney. He told us how to write a letter to resolve the issue. It was as simple as that. We wrote the letter and had our money in less than two weeks. Sometimes I'm a little slow, but I do learn. Since then, I haven't had any legal problems that weren't resolved quickly. My new lawyer is a problem solver.

I'm not sure that putting an attorney on a retainer is a good use of your money. If you write a good contract, your problems should be minimal. However, if you find you're calling your attorney on a regular basis, then you might need to reconsider your business relationship with the attorney. Ask about putting their firm on retainer to represent your company. If the attorney is the problem-solving type, he or she will tell you if it makes good financial sense to do this.

Since your attorney is the person that will help keep your fat out of the fire, so to speak, it's a good idea to cultivate a friendly relationship with him or her. Invite your attorney to company functions, such as Christmas parties or company picnics. If they do a particularly good job for you on some issue, make sure you acknowledge it. You might even send a gift certificate for dinner at a nice restaurant or an overnight stay at a hotel or resort somewhere. Attorneys, like you and me, need appreciation and thanks. I've also found them to be a good source of leads over the years — and you know we can always use another lead!

Writing Your Contract Documents

Many contractors think that they know everything about writing contracts for their work. Unfortunately, I can count on one hand the number of contractors that I've met who really had the ability to write a good contract.

Very few contractors have the discipline to analyze each job they complete to see if their original contract and any additional work orders they had for that job could have been written in a more effective manner. However, this is essential if you're going to develop a good contract on your own. It's the only way you can refine your contract documents into a form that's right for your particular business.

I've found that it's extremely helpful to develop a network of contractors that you can exchange contract ideas with. You can send them a copy of a particular paragraph from a contract and see what they think of the wording. They may be able to improve on it or even suggest an additional paragraph or two to cover some side issues that you left out. A good network can supply you with contract language to cover all types of work and situations. I have a file of contracts from contractors all over the country that I use as references. Most of these are the stock contracts of these contractors. They've made the effort to standardize their contracts to accommodate the type of jobs they do, but even their standard contracts retain some flexibility to cover variations in the work. This takes a lot of thought and study.

I'm not saying that I think you should write all your own contracts. You still need the legal guidance of an attorney. It should be a joint effort. There aren't many attorneys who are experts on construction. They may know what's required in a contract document, but not necessarily all that you need to cover you and your company against the various disasters that seem to haunt our industry. Compile your basic contract language yourself, then have your attorney fine-tune it.

Don't give your contract to your attorney until you've completed it. He or she will need to know everything that's in it in order to help you enforce the agreement. Your attorney should be able to ensure that your customers don't misinterpret your contract — a common cause of disputes. Make sure you have contract language that covers all the general aspects of every job, such as payment schedules, start and completion dates, insurance and bonding, lien rights, etc. Then have optional paragraphs prepared that you can add in or leave out that cover specific items that come up from time to time but aren't essential to every contract. Have everything reviewed by your attorney, and then when you write up a contract, you can pick and choose the paragraphs you need. Be sure you use the exact wording, as approved by your attorney. Don't be creative. Having your contract language reviewed by an attorney is about the best money you'll ever spend.

Using Form Documents

Should you use preprinted forms from a trade association, professional organization or a form contract that you buy at the local stationery store? Be very careful with these! Almost all of the form documents that I've reviewed are far too general to provide good protection for your company if someone's looking to cause you problems. Your contract document must not only cover items required by the law, but it must also provide for payment schedules and other items that are specific to your business.

I've always been amazed by the lack of interest that some trade associations have in helping to compile good contract language for their members. They use the old excuse of not having the authority to get involved in the legal aspects of their members' businesses. The associations have many, many members to draw from. Assembling the best contract language from all their members and making it available for everyone to use in their contracts could hardly put the association at risk. And even if there was some legal risk involved, a disclaimer could eliminate it. Or the association could get a blanket insurance policy to cover that aspect and charge the cost to their members. Having access to a good contract would be well worth the price of the insurance. Since it would probably cut in half the customer complaints that the associations get about their members, you'd think it would be a good venture for everyone involved!

If you're fortunate enough to belong to an association that *does* help its members with contract language, or supplies preprinted contract forms, here's an additional word of caution. Be careful using documents compiled by association architects. They tend to include phrases that dump the responsibility for anything that may go wrong into the lap of the contractor. Any association contract you use should be written by three to five of the general contractors in the association who have been in the business for at least 20 years. They'll know what they're doing and what wording to include to avoid the pitfalls of the business. Make sure the contract language has been approved by a committee and fine-tuned by the association attorney.

Contract Language That Will Keep You Out of Trouble

The very first thing that you should consider having in your contract is a section defining any and all terms that you use in the agreement. You don't want any misunderstanding about terms such as progress payments, substantial completion, material allowance, installed allowance, and so forth.

Customer Involvement on the Job

Once you have your terms clearly defined, there's a statement that I think should be included in every contract you write. It prohibits your customer, the home or business owner, from giving instructions to every person who walks onto the job. In my contracts at Stone Construction Services, or S.C.S., it reads as follows:

> Owner understands and agrees that all communications concerning the job status, job changes, pricing, or any other job issues outlined in this Contract, will only be between the Owner and S.C.S. (job superintendent or principals). S.C.S. will not be held liable for any discussions or agreements made between Owner and any other parties including S.C.S.-hired sub or specialty contractors, S.C.S. suppliers or other S.C.S. employees.

Even when you include this statement in your contracts, you'll still have to remind most of your customers not to discuss the job with your crews or subs when they come onto the site. If you don't, the familiar old line "Oh, while you're here, can you . . ." will continue to come up. You have to really get people's attention on this matter.

It's amazing how some people think that if something is for their job, they can go ahead and buy it and you'll either pay the bill or reimburse them for the purchase. Here's a little paragraph that addresses this problem that's been useful for me over the years.

> S.C.S will not be responsible for any bills, charges, debts, invoices, or other encumbrances incurred in, on, or for this job by anyone other than S.C.S. or its immediate authorized employees.

Additional Work and Change Orders

We'll discuss additional work orders and how to charge for them at some length in the next chapter. However, your customers will need to know about additional work orders before you write up the contract. You'll need to explain them both in person and in your contract so that there's no misunderstanding about how additional work on the contract or changes to the job occur. The more detail you give them up front, the fewer problems you'll have as you go along. Here's how my contracts cover additional work orders:

> Without invalidating this agreement, Owner may order extra work or change the existing Contract by the use of a change or additional work order. A change may consist of additions, deletions, or modifications to the original contract work (the Contract sum and the Contract time being adjusted accordingly), providing the document is mutually agreed to and signed by both the Owner and S.C.S. Such modifications to the original Contract, or subsequent Contracts or change work orders, may only occur with a signed change work order. This change of work order may change the job completion date.
>
> Only one (1) signature from each respective party to this agreement shall be necessary to execute the change order.
>
> Any additional sum shall be paid in full (100%), at the next progress payment due or the final payment due, whichever comes first.
>
> Owner(s) understand(s) a design/estimating and coordination fee of <DOLLAR AMOUNT> per hour will be incurred on the design, drafting and pricing of the change or additional work, whether the change is elected or not by the Owner(s).

S.C.S. will not be liable for any changes made without a completed and signed change work order. S.C.S. will not be liable for any agreements made between Owner and any party(s) other than S.C.S.

Once you've made sure that your clients know and understand that they can't ask your people to make changes without a change order in writing, you'd better be sure that every one of your field people has the same understanding. They should know that nothing, and I do mean *nothing*, gets done on any job unless it's in the original contract or on an additional work or change order signed by the owner. No excuses, no stories, no favors, no changes! They do nothing on the job unless it's in writing and signed by the owner.

Anyone who does work that isn't in the contract, even if the owner agreed to the work, may find that they're not able to collect for that work if it isn't on a change order signed by the owner. If the owner should get upset with you for any reason and decide not to pay you, it'll almost always cost you more to try and collect for the additional work than the actual cost of doing the work itself.

Pets on the Job

An analysis of a kitchen remodel contract for a job that my company recently completed resulted in our adding the following statements to our future contracts:

Owner understands and agrees that any and all animals that may inflict injury on S.C.S. staff or S.C.S. subcontractors or specialty contractors will be kept out of all work areas and all storage areas for the duration of this job. Owner will provide access to all work and storage areas from 7:30 AM to 5:30 PM, Monday through Saturday, for the duration of this job. If at any time access to the work or storage areas is not available to S.C.S. crews, subcontractors or specialty contractors due to the presence of and potential harm from the Owner's pet(s), Owner agrees to reimburse S.C.S. or S.C.S. sub or specialty contractors for expenses incurred for travel and lost time at the rate of $35 per manhour lost, and $.35 per mile per vehicle.

Owner understands and agrees that S.C.S. personnel and S.C.S. subcontractors or specialty contractors will not attempt to enter any work or storage area if Owner's animals have open access to that area.

Owner also understands and agrees that S.C.S. employees and S.C.S. sub or specialty contractors will not be responsible for any pet(s) leaving the home due to doors, windows, gates or other openings in the home being left open due to work in progress.

I decided that we weren't going to lose time or money because of owners who can't control their pets. Neither are we going to tolerate being bitten, chased, clawed, growled at, snapped at, hissed at or rattled at. And, we're not going to baby-sit an animal unless we get paid for it!

Supervision Work

The tough part about doing job supervision work is that you're supplying your knowledge and experience to make a job work, and your abilities are intangible assets that your customer will value less and less as the job progresses. The smoother you make the job run, the lower your value seems, especially when the owner is writing out your checks. Don't expect any compliments to creep in with all that grumbling. Let's take a look at some of the special contract language you need for this type of job.

You need a written contract for this kind of work just as you would for any other job that you take on. There are a couple of things that you must state clearly in your contract. The first thing, once again, is that the owner deals only with you or your superintendent, *and no one else*. Your wording in the agreement might look something like this:

> S.C.S. is solely responsible for securing all labor, materials, subcontractor work and other related items included in Contract, and for scheduling, construction techniques and procedures, and the coordination of all trades and sequences hereunder.
>
> Owner, Owner's agents, or any other parties, are prohibited from directing, or attempting to direct in any way, the progress of the work. They are also prohibited from securing labor, materials, subcontractors or other items that substitute or supplant those included herein unless specifically authorized in writing by S.C.S. Any questions, problems, or requests for changes of work will be directed solely to the S.C.S. job superintendent.
>
> Owner shall be solely responsible to pay any and all subcontractors for work performed at Owner's direction without the written authority of S.C.S., and Owner shall indemnify, defend and hold S.C.S. harmless from loss or liability which results from claims of any subcontractors or others arising from the performance of such work.
>
> In addition, Owner will be solely responsible for all costs resulting from delays or interference on the part of the Owner, Owner's agents, or Owner-solicited subcontractors working on this job. All resulting corrective work, including labor, materials, subcontract or any other costs and construction liens resulting from that work will be the sole responsibility of the Owner at the rate of cost plus 40 percent.

In addition to these restrictions, you should also stipulate that if the customer (Owner) gets involved in any way with a sub, a supplier or any of your employees, they will be 100 percent responsible for anything and everything that happens as a result of that involvement. They will also be required to pay you an additional amount of at least 20 percent of the cost of their involvement as a penalty fee before you move on with the job. The wording might look like this:

Chapter 5: Writing Contracts

> Owner(s) understand(s) and agree(s) not to effect any side arrangements or separate Contracts with any of the employees, vendors, or subcontractors performing work on this job, except as provided by S.C.S. pursuant to the terms of this Contract. Any such agreement must be approved by S.C.S., prior to such agreement or contract, in writing, and the Owner(s) may not hold S.C.S. responsible for the quality of workmanship and materials utilized by these persons. The Owner(s) will also be responsible for any delay caused by the use of outside contractors or other persons.
>
> If Owner enters into such side agreement, without the expressed written agreement of S.C.S., then Owner agrees to pay S.C.S. 20 percent of the cost of such work, prior to the job moving forward from the start date of that work.

You must also clearly spell out in your contract both the total amount that you're to be paid for the supervisory work and the dates the progress payments are due. Include a clause in your contract that specifically states that you'll shut down the job if the payments aren't made on time. That wording might look like this:

> Owner has read, understands, and agrees with the total payment schedule as shown in this agreement. Owner will pay S.C.S. the initial investment, progress payments, and the final payment, as per this agreement and without retention. Final payment of the entire Contract price is due on the day of SUBSTANTIAL COMPLETION of the work and on the issuance of the CERTIFICATE OF OCCUPANCY or by use of the Owner. (Note: See definition of Substantial Completion above.)
>
> If net amount due on progress payment is not paid by the Monday of the week following the due date, S.C.S. reserves the right to stop work until the progress payment has been made, increased by a reasonable sum for the costs of shutdown, delays incurred, and startup.
>
> S.C.S. reserves the right to terminate this agreement altogether if work is stopped for ten (10) continuous calendar days due to failure of the Owner to make prompt progress payments. S.C.S. further reserves the right to recover payment for all work executed and losses from delays or stoppage of the work, including reasonable overhead, profit, and damages. In no case will S.C.S. be entitled to less than their total expenses plus an additional sum of 40 percent of the total expenses incurred. S.C.S. is excluded from all special, indirect, or consequential damages resulting from work shutdown or termination of this agreement.
>
> Payments not made within fifteen (15) days of the due date are delinquent and shall bear interest at the rate of one and one-half percent (1.5 percent) per month, or the maximum amount allowed by law, whichever is more, until paid.
>
> Owner shall pay reasonable costs incurred by S.C.S. in the collection of any delinquent amounts, including attorney fees and costs of preparing and filing liens, regardless of whether suit or action is instituted.

For those who may think that this wording is too strong, or isn't needed, please keep one thing in mind. You're providing a service for which you should be well paid. Your customer will expect you to do a good job, do it on time, keep it clean, and complete the whole process on schedule so that they can get on with their lives. For that they should be willing to pay you on schedule. However, if you don't have that spelled out in your agreement, you could be in trouble. Suppose, for instance, the customer says, "Oh, I have a payment due? I was just leaving for the Bahamas for two weeks, I'll write you a check as soon as I get back!" What do you do? (If you've been in this business more than four or five years, I should say, "What *did* you do?")

If the wording in your contract is specific, you can tell your customer to reread their contract on the way to the airport, particularly the part about what happens if you don't receive your payments on schedule. Calmly advise them that work on the job will stop on Monday. Of course, you should have prepared them in advance with a notice of the due date for that progress payment or for your final payment. This should have been in writing, followed by a second written or verbal notice, to make the due date perfectly clear to those with tendencies toward convenient memory lapses. I hope you're getting the point here. Why leave things to chance? It only takes just a few minutes to prepare a good contract.

Don't give the customer a chance to ruin your day (week or month), and turn a good job into a bad memory, by letting them pay you whenever they get around to it. You don't want to chase after people to get your money, and honestly, most customers don't want that either. You tell them when the payment is due, make it clear, and you'll get your money. You have to worry about yourself — no one else is going to do it for you. Do you think your customers are going to worry about whether you can pay your bills while they're sitting on their assets in the Bahamas? No!

If you have to shut the job down to get paid, the owner should have to pay a penalty in the amount of at least 2.5 percent of the job sales price in order for you to get the job back up and going again. We'll cover the specific wording for this a little later. Be sure you read it carefully. Of course, you'll also want to add in a clause that states that delays created by non-payment shut-downs will automatically extend the completion date of the project by the amount of time it takes to restart and reschedule the job. The owner, under that clause, should also agree to pay any and all bills weekly, with your approval, until the job is completed.

Bounced Checks

Here's a problem that has surfaced on two of our recent jobs. Both customers were nice folks, and we had a great relationship with them. Our jobs went along just as they should, and then we started getting their checks back from the bank. The checks they wrote us for our progress payments bounced! I really can't remember this ever happening before. One couple gave us our check and asked us to sit on it for ten days before we deposited it. We can't operate like that. We now have a new clause in our contract that reads like this:

> Owner understands and agrees that any checks or other medium of payment presented to S.C.S. by the Owner, or Owner's agent, that is returned to S.C.S. for insufficient funds, or any other reason that delays the deposit by S.C.S. of the monies due as outlined below in the payment schedule for this job in this contract document, will incur an additional charge of $50, plus any and all fees assessed by the bank or other institution handling these monies, and any and all other resultant charges, fees or late fees, regardless of the reason or the extent of those charges or fees.

While this may seem like overkill to some, we're not in the banking business and don't intend to finance the jobs for our customers. If they need a loan until next payday or until whoever pays them the money they are owed, let them go to the bank. That's the business banks are in.

Doing Business with Government Agencies

If you are among those who do business with a city, county, parish, state or the federal government, you're probably aware that they usually prepare the contracts for you. Be wary of two very important but deadly clauses that you'll find in most of their contracts: the retainage clause and the penalty clause.

The Retainage Clause — The worst clause that they can include (at least from the point of view of most contractors) is a retainage clause. It would read something like this:

> 10 percent of the contract price will be retained by _____ (city, county, state, etc.) for a period of _____ (30, 60, 90, or heaven forbid, more) days after the final work has been completed.

The document will usually provide some flaky reason why this particular governing body should be able to use your money interest free for the period of time stated. I don't buy into this kind of nonsense, nor should you.

When presented with a clause like that, I respond in the following manner. "I can't participate in any enterprise, whatever the potential profit to my company, which involves a retainage fee. You may specify the conditions for payment in the contract, and base those conditions on the completion of the various stages of the remodeling work according to the specifications. You may also require our company to fix or repair any problems that may arise after we've completed the job, but we can't do the work for you if you withhold any or all of our final payment. If you'll eliminate the retainage clause, we can do business. If not, thank you for calling us, and we wish you well with your endeavor!"

Chapter 5: Writing Contracts

If you take that stand, you must also be willing to walk away from the job. Remember, the people you're dealing with in these situations are bureaucrats, not businessmen. They often have little or no knowledge of how a small business operates.

Sometimes the government entity that you're dealing with has a mandated approach to doing business. By this I mean that there may be a law requiring that the retainage clause be included in all their contracts with private companies. If you decide to work for them (and essentially finance part of their project for them for a period of time), then be sure that you're clear on all the details of how and when you'll be paid for the work. Get a copy of their regulations and make sure you can live with them. Loosely worded edicts that grant discretionary power of payment to some bureaucrat (who may decide to take a four-week vacation right before you're to be paid) can have a devastating financial impact on your company.

The Penalty Clause — The second clause to watch out for is the penalty clause. Again, if it's included in a contract, be very careful. Most penalty clauses in contracts deal with completion dates, or the lack thereof, and read something like this:

> Contractor will be penalized _____ (insert $100, $200, $500, or more) per day for every day that the job remains incomplete beyond _____ (insert completion date).

I've found just one good response to this clause, it reads like this:

> Stone Construction Services shall not be liable for any special, indirect, or consequential damages arising in any manner from delays in performance of the work. No penalties shall be accessed to Stone Construction Services for job completion beyond the date shown above, unless an equivalent sum is guaranteed by the Owner(s) as a bonus for each day the job is completed before the date shown above.

Now some bureaucrats will read that last paragraph and say, "Oh no, we can't do that! That just gives you a reason to cut corners so you can get done earlier!" If you should hear something like that, *just walk away from the job*. Count yourself lucky that you found out what type of person you'd have to deal with before it was too late. Walk quickly and don't look back.

When I started writing this book, I sent a questionnaire out to 90 contractors involved in either new home construction or remodeling. One of the questions on that form asked if they would do business with any company or individual that had either a retainage or a penalty clause in their contract, or required the contractor to write one into his or her contract. All but one responded "No." The one "Yes" respondent said that a penalty clause was OK as long as there was an equivalent bonus clause for early job completion. That question received more additional comments from the contractors than any other question on the questionnaire. There must be a good reason for that, don't you think?

Other Clauses to Watch Out For

One of the biggest gripes that I have with most preprinted contract documents is that they seem to place the responsibility for everything in the lap of the contractor, regardless of who did what to whom and why. Have you ever seen this line in a contract?

> Contractor is solely responsible to confirm all measurements and calculations on all structural members prior to job start.

How about these?

> Contractor is solely responsible for the accuracy of the drawings for this job, and any errors, omissions or discrepancies will be reported to the Owner and Architect prior to job start.

> Contractor bears the sole financial responsibility for any errors, omissions or discrepancies not immediately brought to the attention of the Architect and Owner in writing.

These statements on plans or in the specification book for a given job can be a real time bomb for the contractor who isn't careful. When you are given a set of plans or specs for review and you see these statements, ask the individual giving you the document or blueprints, "Who drew up the plans for this job?" They will normally respond with the name of some individual or company, usually an architect. Then ask, "Did they get paid for those plans?" The answer will be yes. Your next question might be "Why then would I be responsible for their errors or miscalculations?" Very politely hand the contract document or blueprints back to them and ask them to line out and initial those statements, or reassign the responsibility to the individual or company that drew up the plans. If they won't do that, then simply walk away. If you allow yourself to be trapped into signing onto a job where the contract documents include wording that says *you are responsible for everything and the owner and architect are responsible for nothing*, you're asking for big trouble. In all my years in business, I have yet to see a perfect set of plans.

Write the Contract Yourself

Unless you're dealing with a government agency or perhaps a very large company that has a special department that deals with the construction and remodeling of their facilities, you should write your own contracts. If you choose to enter into an agreement where you'll be using another's contract, as in the situations just cited, *have your attorney check their contract over thoroughly before you sign it*.

On any other job, you should write your own contracts. If the building owner is an attorney and wants to write the contract, I tell them, gently but firmly, that I use a very detailed, time-proven contract for all my company's work, and that it has kept us on a sound business footing with our customers for years. I explain that it clearly spells out all the specifications for the job, tells everyone what they can expect from us and, in turn, what we expect from them. As a result, we can build their job with few, if any, problems. Then I ask, "When would you like me to bring the contract by for your review and signature?" If you address the contract issue in this manner, you'll normally find that the problem will go away.

Another approach that I've heard used is this: "Our company has an attorney on retainer, and he writes a very good contract. Why don't we let him write out our agreement? It won't cost us anything, and we know all the legal aspects will be covered. That way we can get on with your job quickly."

If the problem still won't go away, and they insist on writing a contract, then you tell them, "That's fine. I'll write an additional contract specifying the work to be done, and we can jointly sign each other's agreements. When will you have your document ready for my review and signature?" If that's the course they wish to pursue, don't worry. When you write your contract for their signature, be sure and include the following statement:

> Unless it is agreed otherwise in writing between the Owner and Stone Construction Services, it is understood and agreed upon by the party(s) to this agreement, Owner, Owner's Agent(s), Stone Construction Services and all others, that this Contract will supersede any and all prior documents applying to or related to the proposed work as outlined in this Contract, regardless of their date, content or origin.

Now write your normal agreement and present it for signature. If the customer refuses to sign your document with this paragraph included, then I strongly suggest that you look elsewhere for work. There are a few attorneys, just as there are a few contractors, who have the same reputations as the stereotype "used car salesmen." Be wary of getting into business with one of them.

Good Contracts Have Well-Defined Pay Schedules

The following are the first and final pages of our standard proposal/contracts for the work that we do in our construction company. The first page, shown in Figure 5-2, prepares the reader (owner) for the information, specifications and legalese to come. It ties the entire document together and leads the reader into the proposal/contract. The last page, shown in Figure 5-3, describes the payment schedule, confirms that the owner has read and understands what they are signing, and provides for the owner's acceptance of the proposal/contract.

Between the first page and the final page, we spell out exactly what we'll be doing on the job, including all the materials, with the names, brands, model numbers, and anything else that's of importance to the job.

Chapter 5: Writing Contracts

Stone Construction Services
111 Ocean Avenue
Portland, OR 99999
(900) 555-1111

This Agreement For Professional Services is entered into on this _____ day of _____, 199___, by and between STONE CONSTRUCTION SERVICES, an Oregon Corporation, (OR. CCB #100000) (WA. Lic. # STONECS000W) hereinafter called "S.C.S." or "Contractor," and the party(s) signing below, hereinafter called "Owner," governing work to be performed on the property and building located

at: _____
(Address of building to be built or remodeled)

for: _____
(Name(s) of all legal owners of the building)

Phone: _____
(All phone numbers of legal owners, home and business)

S.C.S. shall furnish all labor and materials to perform the work described in the following specifications and attached drawings, and incorporated by reference as part of this Contract and any Addendum attached hereto.

Page 1

Figure 5-2
First page of sample contract

This construction contract is entered into on the _____ day of _____, 19_____, by STONE CONSTRUCTION SERVICES, (OR. C.C.B. # 100000) (WA. Lic. # STONECS000W) hereinafter called Contractor or S.C.S., and the party(s) signing below, hereinafter called Owner. The above specifications, conditions, and job material selection sheets are satisfactory and are hereby accepted. You are authorized to purchase materials and proceed with this job as specified in this proposal. S.C.S. shall furnish all labor and materials to do the work described in the above specifications and Owner agrees to pay S.C.S. as follows:

TOTAL CONTRACT PRICE	$_____
DOWN PAYMENT	$_____
PROGRESS PAYMENT _____	
(Specify payment at start of)	$_____
PROGRESS PAYMENT _____	
(Specify payment at start of)	$_____
(Additional Progress Payments as necessary)	$_____
CASH DUE ON DAY OF ISSUE OF CERTIFICATE OF OCCUPANCY AND SUBSTANTIAL COMPLETION OF THIS JOB	$_____

ATTENTION: S.C.S. will do only that work which is written in the above specifications for the above agreed on amount. The terms and conditions as stated are part of this Contract. This Contract is subject to STONE CONSTRUCTION SERVICES' office approval.

You, the buyer, may cancel this transaction at any time prior to midnight of the third business day after the date of this transaction. See the attached notice of cancellation form for an explanation of this right.

_____   ____/____/____
(Owner's signature)   (Date)

_____   ____/____/____
(Owner's signature)   (Date)

Owner acknowledges receipt of a copy of this Contract, and that they have read, understood and agree with the terms of this Contract and the payment schedule for this job.

_____   ____/____/____
(STONE CONSTRUCTION SERVICES)   (Date)

Figure 5-3
Last page of sample contract

It's extremely important that you plan the progress payments for the start of each segment of the job, never at its completion. You should determine the time and amount of the payments to correspond with your cash flow needs for each portion of the job. As we discussed earlier, you may run across individuals who balk at making a down payment or making progress payments. They want to pay when the work is all done. Don't fall into that trap. It's deadly. Again, a deal is only a deal when money changes hands. If the customer won't put money up to get the job started and agree to the progress payment schedule as outlined on your contract, then I strongly suggest you go find yourself a serious customer.

In some states, there are consumer-protection laws enforced by government agencies that prevent contractors from obtaining down payments. If you're unfortunate enough to be doing business in such a state, then you'll need to put a clause in your contract that'll assure you of prompt payment for your work. You should stipulate that an escrow account, with deposits made for the full price of the job, be set up by the owner. This should be a two-signature account for withdrawal, yours and your customer's. Do not enter into any agreement where the customer's banker or attorney represents the customer.

"If the customer won't put money up to get the job started and agree to the progress payment schedule as outlined on your contract, then I strongly suggest you go find yourself a serious customer."

Make the contract draw dates coincide with the onset of each portion of the job, but with at least two draws made in the first thirty days of the contract period. That way, you can guarantee that you'll have money to pay your bills when your subcontractors' and suppliers' invoices come in at the end of the month. Even though you can't require a down payment, you can try to keep your cash flow in your favor. Make sure that your final draw from the escrow account never exceeds your net profit for the job, so that all your expenses have already been covered.

Subcontractor Contracts

Unless you specialize yourself, you'll probably need to call on the services of a specialist. It's just not economically feasible for a general contractor to try and do their own electrical, plumbing, drywall or other specialty work unless it's on a very small job. So, assuming that at some time you're going to employ specialty contractors on your jobs, you should have a well-written agreement ready for them. There are probably hundreds of such agreements in use today. I've included one here, shown in Figure 5-4, for your review. Again, if you choose to use it, modify it to suit your work and have your attorney fine-tune it for you.

Chapter 5: Writing Contracts

Agreement to Provide Specialty Services to Stone Construction Services

This agreement, made on ____/____/____ by and between Stone Construction Services hereinafter called CONTRACTOR, and _____ hereinafter called SUBCONTRACTOR.

1. The Subcontractor agrees to furnish and install all materials and perform all work necessary to complete the following job: _____

on the project, located at _____

according to the plans or working drawings and specifications of Stone Construction Services and the Owner, and by terms and conditions as the Contractor is bound to the Owner.

2. The Subcontractor agrees to promptly begin said work as soon as notified by the Contractor, and to complete the work as follows: _____

3. The Subcontractor shall cause to be in force Workers' Compensation Insurance, Public Liability Insurance, Property Damage Insurance and/or any other necessary insurance as required by the Owner, Contractor or the State in which this work is performed.

4. The Subcontractor shall pay all Sales Taxes, Old Age Benefit and Unemployment Compensation Taxes due upon the material and labor furnished under this Contract as required by the U.S. Government and the State in which this work is performed.

5. No extra work or changes under this Contract will be recognized or paid for by the Contractor unless agreed to in writing before the work is done or the change made.

6. This Contract shall not be assigned to another by the Subcontractor without first obtaining written permission from the Contractor.

In consideration whereof, the Contractor agrees to pay to the Subcontractor, in _____ payments, the sum of: $_____. The said amount will be paid as follows: _____

Page 1

Figure 5-4
Subcontractor contract

All payments are to be made by Contractor to the Subcontractor after the Subcontractor shall have completed his work as described above to the satisfaction of the Contractor and the Owner.

The Contractor and the Subcontractor for themselves, their successors, executors, administrators and assigns, hereby agree to the full performance of the covenants of this agreement.

_____
(Subcontractor)

_____
(Address)

_____
(City, State & Zip)

_____
(By & Date)

_____
(Subcontractor State License #)

_____
(Subcontractor Tax I.D. #)

Stone Construction Services
111 Ocean Avenue
Portland, OR 99999
(WA. Lic. # STONECS000W)
(OR. Lic. # CCB 100000)
(Tax I.D. # 00-000000)

_____
(By & Date)

Page 2

Figure 5-4
Subcontractor contract (continued)

Be consistent in your use of contracts. Use the same contract for all of the subs and specialists that you hire. Just vary the wording for the job. If it ever comes to a case where you and one of your subs end up in front of a judge, you may be asked to show consistency.

Handyman Contracts

More and more successful contractors around the country are using a Handyman Division and a Handyman Contract for small jobs. The approach to this type of agreement varies from company to company, but the bottom line is that you get paid a profitable rate, with full markup, for work on small jobs. Handled properly, it effectively removes the risk of losing money on these small jobs. That's not to say that a handyman-type job is always small. I've talked with contractors that have run jobs of up to $20,000 or more through their handyman division. If you do work this way, you should have a special agreement to use that governs how you'll be paid.

The following is an agreement that I've used effectively in my construction company for jobs that we estimate at $1500 or less. Some companies raise that limit up to $5000, some only go up to $500 on a Handyman Contract. Think the process through before deciding how you wish to be paid for this type of work. Obviously, you must make your full markup on each job that you do. You could adjust your hourly labor rate higher if you're not comfortable charging cost plus 35 percent for materials, subs and rental equipment.

The contract in Figure 5-5 is a good starting point. Make whatever adjustments you think necessary for the document to fit your company. Have your attorney check it before you use it. Copy it onto your company letterhead (first page only). Make two copies for each job, one for the customer and one for your company.

The Fax

Time for another gut check. Suppose that your prospective customers (husband and wife) work in two different cities, making it nearly impossible to get them together to sign your contract at the same time. You decide to Fax the contract documents to one of them to expedite the process and get on with the job. Do you know the wording you need to use to make your contract legal if signed via Fax? Write it out, or grab a copy of your contract and let's compare it to the Fax wording below.

I know of several contractors, in various states, who've had their contracts voided by the court system because they didn't include the correct wording in their Fax. Even many attorneys aren't aware of the different wording required for contracts by Fax, so do your research. There are very specific rules for contracts signed and enacted by Fax. The wording in the Fax cover sheet shown in Figure 5-6 is from the *Fast Track Proposal Writer*

Proposal for Repair or Replacement Services

The intent of this proposal is to outline the repair or replacement services to be provided by Stone Construction Services (hereafter referred to as S.C.S.) and the estimated cost of those services to the Owner.

The repair or replacement work will be done for:

at:

The work to be done will be as follows:

Owner(s) understand and agree to the following conditions outlined in this proposal:

1. S.C.S. will provide an approximate estimate of the total cost of the project before the job is started. The approximate estimate will include all labor, materials, sub or specialty contractors and rental equipment if needed. This estimate will be an approximation only. Owner understands and agrees to pay any and all actual costs incurred up to the limit stipulated on page 2 of this agreement.

2. Owner will pay S.C.S. for the work as follows:

   A. $44.95 per hour for each S.C.S. employee that works on this job. Time for each employee starts from the time that employee leaves the S.C.S. office and will continue through the time that the employee is at the job site. This time will include all travel to the job site and for material pickup and delivery, job setup, working and cleanup time, travel to dump site and dump site fees as needed to finalize the work.

   B. All materials used on the job, and any and all specialty or subcontractor expenses for this job incurred by S.C.S. will be billed to the Owner at cost plus 35 percent.

   C. An initial payment of 50 percent of the estimated cost of the job will be required from the Owner prior to job start, on any job that will take more than two (2) working days or is estimated in excess of $1,500.

   D. <u>Owner will pay S.C.S. in full on the day of the substantial completion of this job as determined by the S.C.S. job superintendent.</u> If an S.C.S. employee must travel to pick up the payment for this work, then that time will be added to the bill.

_____   ___/___/___
(Owner Initial)   (Date)

Page 1

Figure 5-5
Handyman contract

Proposal for Repair or Replacement Services, Page 2.

E. If S.C.S. is not paid on the day of substantial completion of the job, for any reason, a LIEN will be filed against the property within 24 hours. This will include payment defaults due to NSF Checks or Credit Card Overdraws.

3. Owner is responsible to remove any and all personal or household items in the working area(s), and for keeping all people, pets, personal and household items out of the work areas or any storage areas to be used by S.C.S. during the entire duration of this job. Any delays caused by non-compliance with this paragraph will be billed at the rate as stated in # 1 above.

4. Owner warrants that they have read and agree to this entire proposal, including the payment schedule as outlined below, and that by signing this proposal they accept all the conditions and payment schedule as outlined.

Job Start Date: ____/____/____ Approximate Job Completion Date: ____/____/____

Approximate estimate for work to be done:

Item:

1. _____ $_____.___

2. _____ $_____.___

3. _____ $_____.___

4. _____ $_____.___

Approximate Estimated Total for all Items as listed: $_____.___

Job total not to exceed the following amount
without owner's written agreement: $_____.___

Owner understands and agrees that S.C.S. will do that work, and only that work, as outlined in the item list above for the estimated amount as stated on this proposal. Any additions or deletions to this proposal will only be executed if agreed to in writing between Owner and S.C.S.

_____   ____/____/____
(Owner Signature)   (Date)

_____   ____/____/____
(Owner Signature)   (Date)

_____   ____/____/____
(S.C.S.)   (Date)

Page 2

Figure 5-5
Handyman contract (continued)

Chapter 5: Writing Contracts

Figure 5-6
Cover sheet for Fax documents

Due to the considerations of both time and the separation of the parties to this construction agreement, this Contract is hereby executed by use of an electronic facsimile machine (Fax), and all parties involved agree to the following conditions.

    A. The Faxed document shall be considered a legal and binding offer and agreement between all parties to this Contract.

    B. That this document shall be considered the original counterpart, and that the sender, Stone Construction Services, keep the original and all Faxed documents, and all documents signed and returned on file in Stone Construction Services' place of business. The Owner(s) will be supplied a copy of the contract only upon request, and after authorization of a document stating that the original Faxed document is the original and binding document for this Contract, and is in keeping with the "BEST EVIDENCE RULE."

    C. All documents to this Contract, signed and transmitted via Fax, shall be accepted as legal documents, as will all signatures to this document be considered a legal and binding signature by all parties to this Contract.

    D. This document shall be considered to be legal and binding on all parties to this Contract as of the date of the signatures of the parties on the Faxed documents.

(a software program from Northwest Construction Software) as well as other sources that I've gathered together over a period of several years to cover contracts via Fax.

The cover sheet in Figure 5-6 should be included with your contract if the contract is to be sent, signed and enacted via Fax.

If you live in a state where a notary stamp is required, then you'll also need to include the following:

    E. All parties to this agreement shall sign these documents in the presence of a NOTARY PUBLIC, licensed by the State in which the respective documents are signed. The attending NOTARY PUBLIC will affix their signature and stamp on any and all signature documents transmitted by Fax along with a business address and phone number for said NOTARY PUBLIC.

OK, how did your wording compare? Remember, if you're in doubt about anything, or if the customer is wary of something, put it in the contract. It's always better to be too cautious than to omit something that later becomes important.

Contract Software Programs

Several software companies have come out with programs that you can use to write contracts. Look these programs over very carefully. Don't buy one that simply takes line items from some estimating database, plugs them into a document, and then calls it a contract. Little or no thought has been given to outlining what the document is for and how it's to be interpreted. A contract should contain more than your estimated costs; it should also describe the job. You need a program that defines the terms used in the contract and provides the legalese that's become necessary to protect you from lawsuits. Make sure it contains all the special clauses that we've just reviewed. You want to have some recourse in case your customer interferes with your workers, wants additional work done, or won't pay you on schedule, etc.

There's very little computer software out there that's been designed specifically for construction work. When you buy, take care to get a program written by and for people who work in construction and are familiar with what we need.

This author's company, *Construction Programs & Results*, has created software called *Fast Track Proposal Writer* that will write proposals and contracts for residential and light commercial work. To learn more about this program, go to the CPR Web page at www.markupandprofit.com.

Some Final Thoughts on Contracts

Here's a short checklist of good ideas that will help you make money:

- Write clear, detailed contracts.

- Set a job time schedule and keep to it.

- Get a down payment for the job.

- Set a schedule for progress payments and collect them when they're due.

- Get work change orders signed and dated before the work is done and make sure they're prepaid or paid at the next progress payment.

- Use a job completion punch list (see Figure 5-7) and make sure the customer initials his or her approval for each item on the list before you start work on them, and then signs it when the entire job is completed.

These are the critical areas that create the most problems for contractors. Not following through on these items will add up to delayed jobs, partial payments or no payments at all, and perhaps more opportunities to see courtroom drama in action than you'd like!

Stone Construction Services
111 Ocean Avenue
Portland, Oregon 99999
(900) 555-1111

Final Completion List

Customer: Bob & Mary Jones
Address of Job:
17 N. Suskabush
Washhougal, WA 98671

#	Item	Work to be done	Date complete	Owner approved
10	General conditions			MJ
20	Demolition / tear out			MJ
~~30~~	Excavation			—
~~40~~	Concrete			—
~~50~~	Masonry			—
60	Framing			MJ
~~70~~	Roofing			—
80	Siding			MJ
90	Windows	Replace screen	7-29	
~~100~~	Doors			—
110	Sheet metal			MJ
120	Plumbing			MJ
130	Electrical			MJ
140	H.V.A.C.			MJ
150	Insulation / weatherstripping			MJ
160	Drywall / plaster			MJ
~~170~~	Ceiling tile			—
180	Cabinets			MJ
190	Surfacing			MJ
200	Tile			MJ
210	Floor covering	Reglue base	7-29	
220	Kitchen & bath accessories			MJ
~~230~~	Awning & patio			—
240	Finish carpentry			MJ
250	Hardware & metalwork			MJ
~~260~~	Paneling & fence			—
270	Light fixtures	New bulb @ lav.	7-29	
280	Paint & decor	All by owner		—
290	Debris removal			MJ
~~300~~	Miscellaneous			—
310	Inspections	FINAL	7-30	

Final completion date: ____ / ____ / ____

_____
(Owner)

Figure 5-7
Job completion punch list

Chapter 6
Change Work Orders, Other Forms and Your Markup

Almost every job will have a change to be made, whether minor or major. Changes and additional work happen, that's a given. We're dealing with people and people change their minds. They didn't hear your first description of what you were going to do, didn't pay attention when they read the original contract, and so on. As a result, the job and the contract must change. You'll want to be paid for all the work that you do, including a profit, on both the original contract and the changes. To keep everything recorded, and to simplify the process as much as possible, you must use an Additional Work Order form or a Change Work Order form. Since they are basically the same, from now on we'll just use the term Change Order to describe them.

You can either create your own Change Order form or purchase a preprinted form. If you make up your own, be sure to include all the information that your original contract page included, minus the Rescission Notice, plus a space for the change to be recorded. It's OK to do it this way, but it's a lot easier and less time-consuming if you just buy a hundred preprinted forms.

Do a little research and find a company that produces a good form that suits your type of work. I know that I advised you to stay away from preprinted contract forms because of their generality, but Change Orders don't really fall into that group. If you find one that has all the information that you need on it, and you can get your company name printed on them, go for it.

Keep in mind that you're not in the printing or graphic design business; you're in the construction business. Buy the forms and get on with the business of selling and building your jobs.

Using Change Order Forms

Even the best form is of little use if it's not used correctly. Pay particular attention to the information in this chapter. My company has a strict policy covering the use of Change Orders: If work is done on a job, and it's not covered either by the original contract or a Change Order, then we don't pay for that work. Who pays for it? It doesn't matter, but it won't be our company. All of our employees are made aware of this policy when they start working with us, and they're reminded of it regularly. You should do this too.

If you use a form, whether it's an in-house document or a preprinted document that you've purchased, insist that the field person in charge of the job (superintendent, foreperson, lead carpenter, salesperson, whoever) fill it out completely and turn it in before any work is done. If a Change Order is turned in that's hard to read, mathematically incorrect, or without dates, *signatures*, addresses and a clear description of the work to be done (including measurements, make and model numbers), give it back to whoever wrote it and tell them to complete the document. Make sure you give them a time line in which to get it done. From my own experience, 24 hours is more than enough time to take care of something like this. It's important to follow through on your rules for office and work procedures — it'll save you a lot of money in the long run! Make sure all forms are filled out and used correctly. Figure 6-1 shows one of my forms filled out completely and signed by both myself and the owner. (There's also a blank copy of this form in the back of the book.)

I'm not advocating a blizzard of paperwork that burdens your staff unnecessarily. Keep the forms to a minimum, but you must have a good and accurate paper trail of the work being done on every job. I worked for a company for a while that had a form for everything. It was one of the best-organized companies I'd ever been associated with — from the standpoint of forms and procedures. The problem was that they generated so much paperwork that it took the superintendents almost two hours a day to keep up with it. It took longer for the sales staff to prepare the paperwork for each job they sold than it took them to sell the job itself. The paperwork was excessive and resulted in a very high rate of employee turnover. You can carry a good thing too far — just use some common sense.

Explaining Change Orders to Your Customers

It's very important that you explain to your customers how Change Orders work. You must also tell them when they're expected to pay you for Change Orders and when they can expect to receive a credit for a Change Order. You should do this *before* they sign the original contract. Show the owners a copy of your Change Order form and explain how and when the form is to be used.

Here's how we explain Change Orders to our customers.

For minor changes:

- You come to the superintendent with your change.
- He'll make a sketch, if necessary, of the proposed change.

Chapter 6: Change Work Orders, Other Forms and Your Markup

Stone Construction Services
111 Ocean Avenue
Portland, Oregon 99999
(900) 555-1111

Change Order No. 3

Date:	Owner:
2/10/98	Smith
Job number:	Job phone:
98-1091	—
Original contract number	Original contract date:
	12/26/97

Job name / location:
John & Mary Smith
1111 W. Rose
Vancouver, WA 98000

Change (add or delete) the following work to the original contract:

1. Credit for permits from allowance	<$39.00>
2. Charge above allowance for glass tub enclosure	$65.00
3. Credit for marble deck installation from allowance, 4' deck	<$67.00>
4. Charge for addition of new exhaust fan - light combo unit & switch in main bathroom	$193.00
5. Re-level aluminum patio cover, labor & materials	$230.00

Change the original contract amount by: $382.00
Previous contract amount: $21,872.00
Revised contract amount: $22,254.00

We agree to furnish labor & materials complete in accordance with the above specifications at the price stated above.

_____ 2-10-98
General Contractor   Date

Above additional work to be performed under the same conditions as specified in the original contract unless otherwise stipulated.

_____ 2/10/98
Owner John Smith   Date

Note: This change order becomes part of the original contract.

Figure 6-1
Completed change order for Stone Construction Services

- When it's just the way you want it, he'll compile the estimated cost (or credit) for the change, and bring that information back for your review.
- If you approve the change, you'll be asked to sign the Change Order.
- Minor changes will take a day or two to process.

For major changes:

- Major changes or additions may require our designer, an architect and/or an engineer to rework or add to the original drawings.
- If the plan changes require an expense on our part, we'll include that in our new estimate for the change, plus a percentage to cover the cost of our processing the paperwork.
- We'll prepare an estimate of the cost of the change for you to see. If you still wish to make the change, then we'll ask you to sign the Change Order.
- If for any reason, you decide you don't want to make the change, then we'll continue with the original plan. If we have incurred any expense in preparing the estimate of the cost of the change, you will be obligated to pay for it.
- Once you sign the Change Order, then we can initiate the steps to make the change to the work in progress.
- A major change can take up to two weeks to process.

You need to repeatedly emphasize that no one can make any changes to the original job plan for any reason unless both the customer and you or your company representative sign a Change Order. You also need to tell them that if the change involves an expense on your part to prepare the paperwork for their review, then you'll advise them of that expense up front and give them the opportunity to authorize you to proceed.

We advise our customers how they will pay for Change Orders in this manner:

- Once you sign the Change Order and it's approved by everyone, the additional money for the work is due (*you make the choice here, not the owner*):

  1. Before the work on the change begins.

  2. At the same time your next progress payment is due.

  3. At the time of the final payment for the job as outlined on the original contract. (*NOTE: Unless only the final payment is still due, this is a very bad policy. Change Orders should always be paid either up front, or at the latest, on the next progress payment due.*)

Chapter 6: Change Work Orders, Other Forms and Your Markup

▪ If you have a credit due, we'll deduct that amount from the final amount due on the day of job completion. (*Be sure that the customer understands this. It should also be in your contract documentation.*)

Answer all your customer's questions and make sure they have an absolutely clear understanding of the Change Order process before you have them sign the contract.

No Change Order, No Work!

It's up to you to make sure your superintendent, or whoever represents your company on the job, follows the company policy regarding Change Orders. Whether or not to use a Change Order **should never** be something that's left up to the discretion of any individual. At Stone Construction Services our superintendents know that if they do work that's not on the original contract or not written up on a Change Order, we won't pay them for that work. If they purchase something on their own from a sub or supplier, and the invoice comes into the company from that sub or supplier, the superintendent will be given the invoice and the opportunity to pay it. No exceptions! If they don't want to pay it, then they'd better start looking for another job.

Oh, While You're Here, Would You Mind . . . ?

Heard that before? Sure, we all have. Your response should be automatic, and memorized, "Of course, let me get my sketch pad, and my estimate sheet, and let's take a look at the change that you'd like to make."

The first time you do that with a new customer, you'll get all kinds of reactions, from "Oh, that's too much bother, just forget it." to "This is so small, we won't need a change order, will we?" or "You want to charge me for that? You're already here and it'll only take a minute!" Eventually the responses will sound more like, "Boy, you guys are going to nickel and dime me to death!"

You'll get them all. A successful contractor won't buy into any of that, not for a minute, not even once. Remember: "Of course, let me get my sketch pad, and my estimate sheet, and let's take a look at the change that you'd like to make." That statement should be mentally tattooed across the forehead of every person in your company who's authorized to deal with your customers. If your customers hear that statement every time they ask for something, then you firmly establish the fact that you're going to follow through on your rules about Change Orders.

Being "nice" to a customer and doing all those little extras may or may not get you referrals and more jobs. But even if they did, your profit margin would still be down because of all the unpaid time you spent doing those little extras. And for those who think they can arbitrarily violate this rule, even once, and then go back to the proper way of handling changes on

Markup & Profit: A Contractor's Guide

that job, watch out! Once you've opened Pandora's box, you'll never get it closed again. It's so much easier, and oh so much cheaper, to do things the right way *all the time*.

Be Consistent

Let me tell you a short horror story that I heard from a contractor I know. He had contracted with a couple to do some work on their home. The contract had all the right clauses in it about payment schedules, change orders and arbitration.

When the job was about 60 percent done, the customer requested an alteration in the design of a closet being built in the daylight basement. The superintendent made lots of notes, especially in regard to the changes the customer wanted in the electrical system. The agreement was that they would start work on the changes that day, and the superintendent would leave a Change Order covering the alterations for the owner to sign when they got home. The superintendent couldn't give an exact price for the electrical work because the electrician couldn't come out until the next day to give the company a firm price for the changes. The superintendent "guessed" at the amount for the additional electrical work and the couple, not wanting to hold up the work, assured him they would pay for the changes when the invoice came in.

I'm sure you're already miles ahead of me. The changes were made, and when the invoice came for the electrical work, it was double the amount that the superintendent had "guessed" it would be. The couple refused to pay for the additional electrical work. Then they started finding all kinds of things wrong with the job, complained to every association, BBB, State Contractor's Board and anyone else they could get to listen to them. They even accused the contractor of threatening them when the contractor said he would start legal action if they didn't pay.

"There should be nothing left to chance when it comes to your company employees handling Change Orders."

The contractor sent his standard letter. It said that if the owners refused to pay the amount owned, he would put a lien on their property, start arbitration as called for in their contract, and then start lawsuit proceedings if necessary. The owners went to their attorney, who promptly wrote a letter to the contractor that said, "You did the electrical work without getting a signed Change Order that included the additional cost to your customer. That's in direct violation of the contract that you wrote. The owners aren't going to pay for the change in the electrical work."

The contractor's attorney basically told him the same thing (except he charged him $150 to tell him): "You wrote the contract, you violated it. Your chances of collecting the additional $400 are slim to none. Clean up your act and abide by your own contracts. Don't do work that's not on your contract, or on a signed Change Order." That was a $550 lesson that he won't soon forget. And I have a lot more stories like this one — many with higher price tags!

It's been my experience that most customers keep their word and pay for changes that they've asked for and agreed to pay for. Every once in a while, however, you run across that less-than-honest person who'll spot a loophole in your operation and take advantage of it. You must be diligent about following the procedures that you set up for work changes on your jobs. Review the process with your employees at least twice each year. You could even take it one step further and put your Change Order procedures in writing and have every employee read it and sign it. That way you won't have any disagreements with your own people about how changes are to be made. There should be nothing left to chance when it comes to your company employees handling Change Orders.

Sales Commissions on Change Orders

If you have salespeople selling jobs on a straight commission basis (which is the only way a salesperson should be paid, especially in remodeling work), then you must set a maximum limit on the amount that you or your superintendent can write a Change Order for without including a sales commission for the salesperson who wrote the original contract. Use a uniform figure on all jobs, something like $500 or $1,000. So, when the dollar amount of a change reaches this sum, the superintendent will know to ask the salesperson to come out and write up the Change Order so that the salesperson will get his or her commission. Your superintendents need to have this issue very clear. They have enough to worry about without the added concern of having to decide at what point they need to get a salesperson back out to write up a Change Order.

It should also be mandatory that if a superintendent needs to have a salesperson write up a change, that salesperson should be required to get back to the customer within 24 hours to effect that change. If they don't, and the superintendent has to write up the change order, then the salesperson should forfeit the right to any commission from that Change Order, regardless of the size (dollars) of the change. Salespeople tend to find all kinds of reasons why they can't do something, especially if they think they can pawn their work off onto someone else. You need to give them a monetary incentive to keep on top of their jobs. If they don't write up the change, then the commission should go to the superintendent who does.

Markup On Change Orders

Should you include markup for Change Orders? If so, how much? What kind of markup should you use? Questions like these have haunted contractors for years.

To me it's simple. You should charge at least as much for Change Orders as for any other work that you do. Apply the same markup to a Change Order as you do on the original contract, if not more. Yes, I do think there are times and situations when a higher markup is justified. Expenses for changes in the midst of construction are often higher than they'd be if the same task were incorporated into the normal progression of the job. They often add more pressure to the job as well.

For example, what happens if you have an exterior wall already framed up and in place, with door and window openings where the plans specify, and the customer decides that he'd like to replace the door and one window with an 8-foot sliding glass door? And, while you're at it, he'd also like you to add a hose bibb on the right corner of the exterior wall. You know it's going to cost more to make that change than if you'd hadn't already framed for the other openings. You not only have to go through the process of drawing up and estimating the change, but you'll also have to make adjustments in your schedule. You're going to end up with a job delay due to the time required for a new plumbing permit, scheduling a plumbing sub, ordering and picking up materials, and reframing the wall.

In this instance, should you increase your markup by an additional 5 to 10 percent? What do you think? That seems fair to me, but every case is a little different. For you, the extra trip to the city (county, parish) permit department, with the time, expense, and headache of dealing with bureaucrats, might be worth a 50 percent markup, just for the frustration — but you can't charge for that! You have to be careful how you present a Change Order for some items. There are times when you'll be requested to make a change, and when you arrive at the final numbers for that work, it'll seem high — even to you. You must maintain your credibility with your customer. Charging them a high price for what seems like a minor change is often hard for them to take, unless of course you can clearly justify the cost.

The Right Percentage

There's a certain amount of judgment that you must use when you establish the price of any changes that come up. If you've been in this business a while, you know there are many things that can affect the financial outcome of a job. The math we discussed earlier in the book should normally dictate your correct *minimum* markup. However, many of the good contractors I've talked with over the years believe that on Change Orders your markup should be higher than on the original job. There are several reasons for this, but the main reason is to cover those unanticipated things that can, and will, go wrong on any given job. If you raise your markup on changes, you can call it a contingency factor; if you don't raise your markup, you can call it Murphy's revenge! Some contractors build a contingency factor into their job costs on Change Orders, while others increase their markup.

Probably the best way to do this on Change Orders is to raise your markup. Many companies have a firm policy that if the hard cost of the change is between $1 and say, $3,000, their markup automatically goes up a predetermined percentage. For example, on a change with job costs of $1,000, you would up your markup from 1.5 to 1.65 or maybe higher.

Be careful how you set your increased markup for smaller jobs. You must be clear with your salespeople, especially new people, exactly how you expect them to apply the added markup percentage. Suppose you normally used a 1.5 markup on your job costs to arrive at your sales price. You then tell your sales staff to add an additional 25 percent to the markup on any change with hard costs of $2,000 or less. This is what can happen.

- Salesperson Kelly has estimated a job change at exactly $2,000. What should her new markup be? Multiplying $2,000 by 1.50 (the normal markup) she gets a price of $3,000. If Kelly increases her markup from 1.50 to 1.75, which is the way you want your staff to interpret a 25 percent increase in markup, then the new price would be $2,000 × 1.75 = $3,500. That's exactly what she does. You have successfully communicated to her the way you want your markup calculated.

However, not every member of your sales staff *heard* the same thing. Well, they all heard the same thing; they just didn't *understand* the same thing. To be fair, there are a couple of other ways you could add 25 percent to your markup. That's why it's important that everyone understand how you want the percentage figured.

- Salesperson Smith also has a job that comes in at $2,000. He thinks, OK, if I must increase our markup by 25 percent then I should multiply 1.50 by 1.25. That gives me a 1.875 markup. So he multiplies $2,000 by 1.875 and gets a price of $3,750. Those are good numbers, but that's not how you meant to have the markup calculated.

- Salesperson Jones is also good at math and takes a little different approach. He takes the job costs of $2,000 and uses the normal markup of 1.50 to get a figure of $3,000. Then he multiplies that by 25 percent to get $750 ($3,000 × .25 = $750), which he adds to the $3,000 price to come up with a new price of $3,750. Again good numbers; but again not what you meant.

- Now we come to salesperson Miller, who's always trying to figure out how to get the price lower. He thinks it will help him "sell" the job or justify the cost of the change order. Believe me, every sales staff has at least one guy like this. He can add a twist to the calculations that no one else would have ever thought of! Miller also has a job with costs of $2,000. He marks it up at the usual 1.5 figure to get $3,000. Then he simply adds 25 percent of the $1,000 markup to get his final figure of $3,250 ($1000 × .25 = $250. $3000 + $250 = $3,250). That's not what you had in mind at all! But he'll tell you that's just what you said to do.

The number of ways you can figure these markup scenarios is limited only by the creativity of the salespeople you have working for you. If you don't make it crystal clear to everyone how to use the markup factor correctly, they'll invent a whole new math procedure for your company — and always with the best intentions. Write it out for them. Put an example on a blackboard or an overhead projector and work the problem through. Make sure there's no way they can misinterpret how you want the percentage figured.

Justifying Your Markup

Now, how do they justify this cost to the customer? Suppose you tell your staff that on any jobs with job costs up to $2,000, you want them to use a markup of 2.0. The next day a customer asks Ms. Kelly for a new bathroom door, which she'll have to write up on a Change Order. The existing door is warped and will have to be removed. Kelly estimates the cost of the new door and trim at $90, adds $45 for labor to remove the existing door and install the new door (1½ hours at $30 per hour), plus a $10 dump fee, and arrives at a cost of $145. She then applies the 2.0 markup to the cost to arrive at a Change Order price of $290 for the new door.

That seems like a lot when you can go down to the local discount home-improvement store and buy a similar door for $59.95. How can Kelly justify her price? Here's a way that works for me.

You tell the customer that the cost of the new bath door, installed, will be $440 — including finishing the door, either with paint or stain and varnish, to match the rest of the doors. Then you tell them that if they want to finish the door themselves, the cost will only be $290. Would they consider finishing the door to save on the cost?

Here, you're using a minor decision to get to a major decision. You'll be amazed by how many times they'll say, "No, you go ahead and finish it." The next thing out of your mouth should be, "OK, give me a minute to write out the Change Order and we can all sign it and get your new door in place."

Countering Objections

Sometimes they'll respond with "How come that door costs so much? We looked at one down at the discount store last week and it was only $59.95!" Your answer should be: "I saw the same door when I was in that store last week. It's similar, but it isn't a door you would want in your home. Besides, when you buy a door at the discount store, the door is all you get. It doesn't include the lockset, the jamb and casing or the cost of pickup and delivery. You still have to tear out and dispose of the old door and then pay someone to install the new jamb and casing and hang the door level and plumb if you want it to open and close properly. There's also the cost of setting the lockset, and applying the finish to the door so that it matches the rest of the doors in your home. When we install the door, we do all that for you, including cleaning up the mess and hauling the old door and trim to the dump, and paying to dump it. We'll have at least two, maybe three people working to get the job done quickly and done right. What would do you think a fair price is for that job?" More often than not, if you've given the customer a good explanation of the costs involved in installing the door, you'll get the job. They'll usually say something like, "OK, it sounds reasonable when you explain it that way. Let's do it."

If they continue to squawk, you have to remain strong and know to stop the discussion at the point where it could turn into an argument. *You never want to debate your costs with a customer.* Just tell your customer very simply, "We can't install a new door for you if it's going to cost us more than you want to pay. Why don't you think about it and get back to me on the change. You don't have to make a decision right now. If you decide you want us to do the job, we'll write the Change Order then, and get the new door in place for you. If you continue to feel that this is more than you want to invest in a new door, then we'll stick to our contract and complete the job just as we'd originally planned." Then let the subject drop, move onto something else and continue your job. If they change their minds about the door, they'll let you know.

Two things are very important here. *First, never cave in when the customer squawks about the price.* If you do, even once, they'll continue to squawk about every price you quote for everything else on the job because then they'll know they can get you to bend. And, what's worse, when you tell them their final payment is due, they'll squawk about that too! And who'll be to blame?

And second, when you're dealing with a customer who takes issue with a price change, *never allow the customer to talk themselves into a hard, fast position that your price is too high.* If you do that, they'll feel they must save face, and the additional work then becomes a dead issue. After that, they'll be on the lookout for other ways to challenge you as the job moves along. Rest assured, for every ounce of argument they feel they've lost to you or your staff, they *will* extract several pounds of flesh to even the score. If you find that you've backed someone into a position or a situation like that, then it's best to give him or her the new door, and be happy you got out as cheaply as you did. Learn from the mistake, and next time don't let it get to the point of an argument or face-saving situation for the customer.

You must give your customers a good explanation of the work that you'll be doing on any change. They need to feel satisfied that you're offering good value for your services so they don't think they have to argue about the price. By the same token, you must get enough money for every job to cover your job costs and your overhead, and still make a profit.

Other Forms

To make money in this business, you must get your jobs started, completed and collected on time. A good set of forms will go a long way in accomplishing that task. You should have all your procedures and the forms needed to implement those procedures in your Method of Operation Manual. Let's look at some of the other forms you need.

Lead Slips

A lead slip is the very first form you'll need. It's the form you grab when the phone rings. It should list all the information that you need to get from a potential customer so that you won't have to stop and think about the conversation. Anyone who answers the phone in your company should be thoroughly schooled in how to take a lead for your sales staff to follow through on. Figure 6-2 shows the lead slip, or what we call our Quotation Request, that we use at Stone Construction Services.

Sales Forms

When the lead has been set, you need a form on which to analyze and estimate the projected job. I use the Unit Cost Estimating Form that we looked at back in Chapter 2 (Figure 2-2). I fill it out and make notes as I discuss the job with the customer. Then, when I return to my office, I transfer the information to my computer estimating program and complete the estimate.

You should also have a form that you and your sales staff use to set and check monthly sales goals. It should include the number of leads each salesperson took, the disposition of each lead, their sales projections and the commissions they earned. Figure 6-3 shows the Monthly Sales Report that each salesperson on our staff must turn in at the end of each month. If you use a form like this, you can easily calculate commissions, review goals and set sales projections for your company.

When a job is sold, you need a pre-job conference form that lists all the people who need to come out to the job site for the layout (or pre-job conference) so that you can be sure they've been contacted and will be there. This should include the salesperson, job superintendent, all the subs who'll be involved in the job, and your customers. Figure 6-4 shows a Layout Sheet for S.C.S. Construction.

And of course, no sale can be completed without a written contract. You need sales contracts, any required state forms, and the federally required Right of Rescission form. We looked at these forms in the last chapter.

Quotation Request

Stone Construction Services
111 Ocean Avenue
Portland, Oregon 99999
(900) 555-1111

Date:	Request taken by:

Customer name:

Job address:

Mailing address:

Home phone: ( ) _____	Best time to call: _____ AM/PM
Home fax: ( ) _____	His pager: ( ) _____
His work: ( ) _____	His mobile: ( ) _____
His fax: ( ) _____	Her pager: ( ) _____
Her work: ( ) _____	Her mobile: ( ) _____
Her fax: ( ) _____	Referred by: _____

Job

Major remodel	Room Addition	Kitchen	Bath
Sunroom	Dormer	Basement	Siding
Deck / Exterior	Handyman	New home	Commercial

_____
_____
_____
_____
_____
_____
_____

Other information: _____
_____

Appointment time: _____ AM/PM   Day: _____   Date: ___/___/___

Directions to job site:
_____
_____
_____
_____
_____

Referred to: _____

Figure 6-2
Lead slip

Chapter 6: Change Work Orders, Other Forms and Your Markup

Stone Construction Services
111 Ocean Avenue
Portland, Oregon 99999
(900) 555-1111

Monthly Sales Report

Month:	May, 1998
Salesperson:	Jones

Date	Customer	Callbacks	Quote	Sale Yes / No / Pending	Notes
5/1	Smith	II	$7,749	(Y) N P	bath & misc
5/4	Emmich	I	$11,880	Y (N) P	dead, went w/brother-inlaw
5/6	Veroske	IIII	$37,228	Y N (P)	waiting on financing
5/7	Knaub	II	$1,590	(Y) N P	misc. in kitchen
5/9	Thompson	NI I	$84,636	Y (N) P	dead, will move
5/12	Robinson		No quote	Y (N) P	legal problems w/former owner
5/15	Gruhar	II	$21,448	Y N (P)	waiting on bonus check
5/18	Rice	I	$3,309	Y (N) P	dead, will do himself
5/20	Watkins		—	Y (N) P	owner not home for appointment
5/21	Boyle		$±250	Y N (P)	they will "think about it"
5/26	Long	II	$12,455	(Y) N P	windows, deck
5/27	Mitchell	I	$6,729	(Y) N P	partial kitchen
5/28	Masuda	I	$2,988	Y (N) P	dead, will wait
5/30	Goldfarb	III	$46,330	Y N (P)	
5/30	Vu			Y N (P)	working, unable to contact
				Y N P	
				Y N P	
				Y N P	
				Y N P	
				Y N P	
	Total:	25	$236,592	4 6 5	

New leads: 15
Leads contacted: 13
Sales: 4
Pending: 5

Ratio of leads to sales: 4/13 or 1:3.25
Total sales for month: $28,523
Commission earned: $2,852

Figure 6-3
Monthly sales report

Stone Construction Services
111 Ocean Avenue
Portland, Oregon 99999
(900) 555-1111

Layout Sheet

Layout date: 7/7/98	Sale date: 6/29/98
Estimate by: MCS	Plans: Yes
Customer: Bob & Mary Jones	
Phone: 555-661-1700	Job phone: 555-831-1710
Owner at layout: (Mr) / Mrs / Ms	Jones
Job superintendent: Bill	

Job address:
17 N. Suskabush
Washougal, WA 98671

Trade	Subcontractor	Date called	Meeting time	OK / By
Permits	City - plbg - elect			
Demolition	SCS			
~~Excavation~~				
~~Concrete~~				
Framing	SCS			
~~Roofing~~				
Siding	SCS			
Windows	PBP - ordered			
~~Doors~~				
~~Gutters / downspouts~~				
Plumbing	Best	7/2	10:15 a.m.	LC
Electrical	Willmett	7/3	10:45 a.m.	Robin
H.V.A.C.	No			
Insulation	SCS			
Drywall	Hansen	7/3	11:15 a.m.	Keith
Cabinets	Andy's	7/3	11:30 a.m.	Joe
Surfacing	Superior	7/3	11:45 a.m.	Cliff
Floor cover	Superior	7/3	11:45 a.m.	Cliff
~~Fencing~~				
Light fixtures	By owner - SCS install			
Painting	By owner			
Debris removal	SCS			
~~Lumber~~				
Finish materials	SCS			

Job Checklist

Review contract	√	Schedule made	√
Review plan	√	Start date	7/13
Selections completed		Permit	
Salvage items		Job sign	
Pets	None		

Directions to job site:
West on Highland to Dripwater, right
to Suskabush, right, 3rd home on right.

Figure 6-4
Layout sheet

Checklists

An in-house checklist for each job sold is also a good idea (see Figure 6-5). We have ours printed on the job folder and it helps to make sure that the job paperwork gets to where it's supposed to be on time so that the job progresses on schedule.

Work Schedules

A production or job schedule must be drawn up for every job, reviewed with everyone that will be involved, and agreements reached on what work will be done by whom, and when. We use a simple work schedule sheet that shows what day each item on a particular job is to be done (see Figure 6-6). Anyone can look at this schedule and see at a glance where we are with the job, and what subcontractors we're expecting or work we're doing on any given day. A good schedule ensures that materials are ordered, plans and permits are secured, and the cash flow keeps up with the flow of the job. The cash-generating ability of each job should also be included in the cash-flow program for the company.

All this work can be done much easier, and far quicker, with the proper forms. Again, don't get carried away with too many forms. Just be sure you have a good paper trail to keep track of each job. There are blank copies of the forms we've just discussed in the back of the book. You can use them as they are, or modify them to suit your own business.

Computer-Generated Forms

If you're not using computers in your business, you're wasting time and effort, not to mention a lot of money. You can generate almost all the forms you need on your computer, and also track a lot of information on computer files that would otherwise require a form. A good computer is one of best tools that you can purchase for your company.

There are some company owners who try to operate out of their head. They just won't let go of any part of their business. They don't want to have to deal with forms. They think they can remember everything and keep track of everything in their heads. They think that keeps them in control. This is the least efficient means of running a company that I can even imagine. Get your forms. Make sure everyone knows what forms you want him or her to use and how they are to use them. Then, let them do their jobs.

Stone Construction Services
111 Ocean Avenue
Portland, Oregon 99999
(900) 555-1111

New Job Checklist

Job sale date:	Job #:
Salesperson:	Job superintendent:
Contract amount:	Allowance amounts:

Customer name / address:

Manager review:
- Contract & job # _____
- Down payment _____
- Allowances _____
- Selections _____
- Estimate _____
- Layout sheet _____
- Specialty items _____
- Thank you letter _____

_____
(Manager OK)

Bookkeeping:
- Check all math _____
- Check pay schedule _____
- Job set up _____

_____
(Bookkeeping OK)

Job superintendent:
- Read contract _____
- Check estimate _____
- Check allowances _____
- Check selections _____
- Check special orders _____
- Check layout sheet _____
- Confirm subs at layout _____
- Schedule permit appointment _____
- Schedule plans _____
- Confirm start date _____

_____
(Superintendent OK)

Salesperson:
- Thank you note _____
- Confirm layout date w/owner _____
- Check all forms complete _____

_____
(Sales OK)

Projected job start date: _____ / _____ / _____

Projected job completion date: _____ / _____ / _____

Figure 6-5
New job checklist

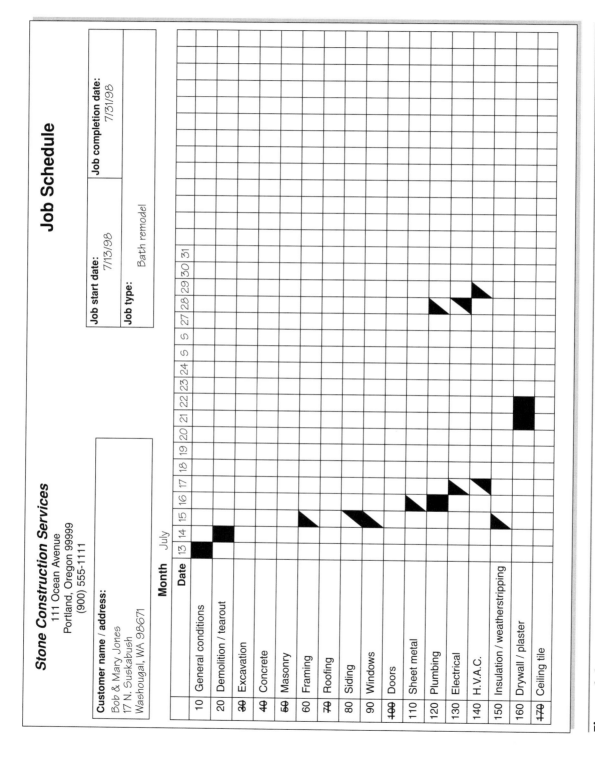

Figure 6-6
Job schedule sheet

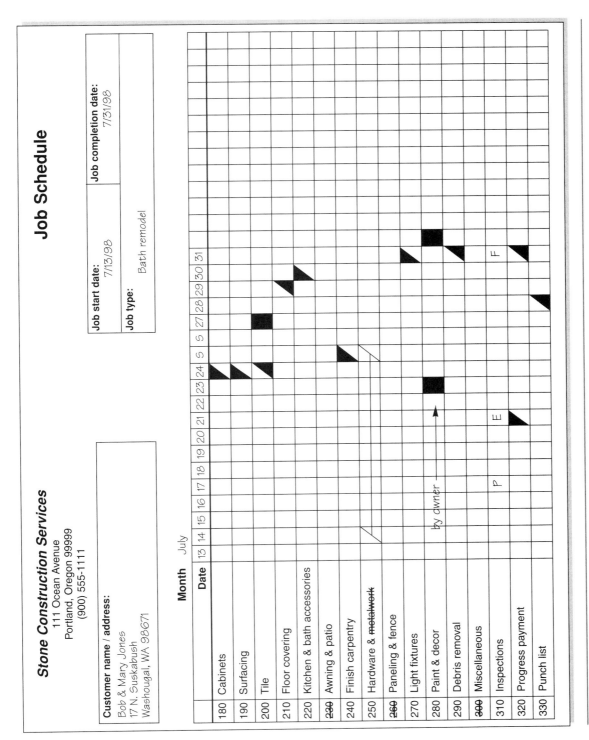

Figure 6-6 (continued)
Job schedule sheet

Chapter 7

The Mathematics of Your Business

While this might come as a shock to some of you, contractors are often guilty of spending more money than they earn. I know this is true because I have a tendency to wander down that path on occasion, especially when I get a notice about an auction in the mail! The end result of this problem is that many contractors have more bills coming in at the end of the month than money to pay them. It doesn't take long for a situation like that to put you out of business. OK, so now we know it's a problem, how do we deal with it?

In my opinion, 99.9 percent of all contractors have no business with the company checkbook, nor should they be the final decision-maker when it comes to spending the company funds. Most of you will disregard my opinion, but even so, I hope you'll follow my next advice. If you recognize that you have a spending problem, put yourself on a budget.

Sounds simple enough. Make a budget and stick to it. Sure, when pigs fly! You ask how I can expect you to make a budget if you don't know how much business you'll have coming in? It isn't easy. But nothing about running a business is really easy. If it were easy, every schoolteacher and fireman from Maine to Oregon and from Canada all the way down to the Mexican border would be doing it. It takes time, effort, work and a lot of projections to create a workable budget. You need to keep accurate records and do a good analysis of each job that you do. And, if you have the discipline and energy to do the work and *set a budget*, the hardest part will be to *follow it*!

You must sell enough jobs to match the volume your numbers are predicated on. If you do that, the end result will be a nice profit by the end of the period. That's something many contractors seldom see!

Setting and Keeping a Budget

Let's talk about your budget, how to set it up and how to follow it. We'll start with this advice: *IF IT'S NOT IN THE BUDGET, YOU DON'T SPEND IT!* It doesn't make any difference how bad you want something; if you don't have the money, you don't buy it. If you really needed it, you would have included it in your budget. Since you didn't, for now you'll have to rent that piece of equipment or do without that new truck. Next year you'll remember to put it in the budget. That's the kind of discipline it takes to make money.

Cash Flow and Adjustments

The most important calculation you need to set up a budget is a cash-flow projection for your company. Construction projects have a variety of time schedules. Remodeling projects can range from three weeks to six months or longer and new home construction can take from seven weeks to a year or more. Specialty work can be as short as a few hours one day with the final a week or two later, to a year or more in length. A commercial project can take a week or last well over a year. So you see, there are a variety of time schedules involved in the work that we do. One schedule won't work for everyone, or even every job.

A prudent contractor will always keep a sharp eye on the money coming in and going out. Eugene Peterson, owner of C.A.R.E., a casualty repair company in Salt Lake City (and I think one of the best contractors in the U.S.), once gave me some advice. He said "If I were to list one thing and one thing only that a contractor should focus on to survive in the construction business, it would be **cash flow**!" We would all do well to heed his advice.

The biggest expense for most contractors is usually labor. Labor normally takes up 45 to 65 percent of any given job cost. On top of individual project labor costs, those of us with regular employees also have a payroll to meet every week or two. How and when you pay your subs is one of those items that's determined by how you run your business, and your relationship with your subcontractors.

So, how can you as a contractor make the predictions necessary to establish your cash flow needs? What steps can you to take to ensure that the money you need to pay your bills and meet your payroll comes into the company on time? And, what do you do if for some reason that money doesn't come into the company on time? You'll still need to pay your bills.

Before we discuss these questions, I want to mention something you rarely hear the "industry experts" talk about. It's an *Operating Capital Reserve Account*. This is the money you set aside to tide you over when the expected or due money doesn't come in on time or when you have taxes due (sales or payroll) and no other money saved up to pay them.

So, in addition to predicting your cash flow needs, you also need to develop a method of setting up and maintaining an Operating Capital Reserve Account, or O.C.R.A. as we'll refer to it from now on. An O.C.R.A. is your best solution for dealing with the type of cash flow problems we experience in our business. It'll give you the ability to keep your bills and taxes paid when there's little or no money coming in.

And here's a little side note, while we're on the subject of taxes. A supervisor at the Washington State Department of Revenue told me that of all the businesses that operate in the state of Washington, there are far more tax liens filed against contracting companies than any other type of business. They make up 20 to 21 percent of the total liens filed. Because of this, the Revenue Department watches closely to make sure contractors file their taxes on time. I'm sure this isn't just a localized problem, so keep this in mind when your taxes come due.

As most of you know, in our line of business things don't always go as planned or happen just when they're supposed to. You continually have to do some refiguring here, recalculating there and reshuffling someplace else. Does the phrase "tap dancing" come to mind? Well, that's about how it works. A lot of contractors tap dance around their finances. You start essentially with your best guess, and then you fine-tune your procedure until you get it down to a workable formula for your company. The basic procedure that I'm going to outline here will work for *everyone* who's reading this book. The final numbers and percentages will be different for each company, just as markup is different for each company, but the numbers will work. The important thing is that you do it.

Establishing Cash Flow Predictions

There are a number of steps you need to take to establish your cash flow predictions. Back in Chapter 3 we talked about projecting your annual Dollar Volume Built, and using that to estimate your Overhead Expenses and Profit. You've already established these amounts for your company, or should have if you've put out the effort to get this far into this book. For this example, I'm going to use the same numbers that we used in Chapter 3 when I showed you how to establish the markup for your company. Our example was a remodeling company with a first year dollar volume built of $150,000, and overhead expenses estimated at about 25 percent, or $37,500. Job costs and profit don't come into the mix in this problem. So, this is what we have:

- Projected Dollar Volume Built   = $150,000
- Overhead Expenses at 25 percent   = $37,500

Let me remind you once again that the numbers we use for our examples don't make any difference. It's the method or formula that we're interested in, and the end result that counts.

Next, we take our annual overhead figure and divide it by 12 to get our approximate monthly overhead:

$37,500 ÷ 12 = $3,125

Now common sense will tell you that some months our overhead will be more than $3,125 and some months it'll be less. For example, quarterly taxes will increase our expenses during four of the twelve months in the year. However, we know that on the first of each month we need to have approximately $3,125 to cover the checks we'll be writing during the next thirty days. By the same token, we know that we must sell, build and collect a minimum of $12,500 ($150,000 ÷ 12 = $12,500) worth of work prior to that thirty-day period to have that $3,125 available to pay our monthly overhead expenses.

One of the keys to making this process work, as you'll soon see, is knowing how far in advance we need to have work sold to assure that the money will flow into our company in time keep up with our expenses. Let's take a look at what we can do to make sure that our sample company generates the cash flow that we need.

"The reality is that we need to develop a certain number of leads each month so that we can keep jobs coming in at the rate of at least two, and hopefully more, each and every month."

We've established that we need to sell, build and collect $12,500 a month to realize the $3,125 to pay our overhead expenses. We need to complete this process at least one month prior to the month in which the overhead expense we're covering will occur. This is where most companies run into problems. They wait until they have the bills in hand and then try to figure out how to raise the money to pay them. They're always playing catch-up, and you can certainly predict the ending to that game.

Let's assume that our company can build a $12,500 job in five weeks. Yes, you can do that size job in one week, but I've also seen the same size job take twelve or more weeks to build. The important thing is to know your own time line and habits so you can predict when you'll get the job built. For our purposes, we're going to use five weeks as the construction time needed for this job. Then, working backwards, we know we'll have at least two weeks from the date of our contract to the actual job start, bringing our total up to seven weeks for this job. Then we need to add another two weeks back to the date of the initial call on the customer to finalize the design, complete the estimate, write the contract and allow time for the three-day "Right of Rescission" to expire. That gives us a new total from the first date of our customer contact through job completion and getting the final check in our account of at least nine weeks. Remember, it's not just how much you sell and when, but also how much you sell, build and collect *and when you collect on it*. That's the key to this process! To be on the safe side, and allow for any problems that may come up collecting or receiving payment, we should add at least one week more to the nine weeks we have established, giving us a grand total of ten weeks for the job.

So to pay our bills in June, we must have sold the necessary volume during the last week of February and collected the money due for the work by the end of April (approximately ten weeks later). That way we're guaranteed to have the cash in the bank at the first of the month to cover our checks as we need to write them. Now remember, this one sale of $12,500 is to cover one whole month of overhead. We all know that jobs come in sporadically and that they'll all be different sizes (dollar amounts), so how do we balance this out so that it works?

To sell the appropriate amount of volume, at the times needed, we simply have to work the plan backwards. If you know anything about sales, you've probably already done most of the math that I'm about to cover. However, to be sure we're all on the same page, let's review.

First we need to calculate our average job size. Let's say that it's $6,250. That means that to reach our goal of $12,500 in sales, we must make two sales each month. To make two sales, using a ratio of one sale for every five leads that we go on, we must then generate ten leads. (Here of course, you'll need to plug your own numbers into the formula to get the ratio that's appropriate to your company.) To generate ten leads we have to get the phone to ring at least ten times, and probably more like fourteen or fifteen times in order to weed out the tire kickers. Keep in mind now that we probably can't get ten leads from those fifteen calls in one day and also make two sales. That just doesn't happen. The process needs to be spread out over a period of at least two or three weeks. The reality is that we need to develop a certain number of leads each month so that we can keep jobs coming in at the rate of at least two, and hopefully more, each and every month. If we can do that, then three months from now we'll have the money to pay the bills that we already know will be coming in.

We also have to remember that it takes a potential customer a certain amount of time after their first contact with our company via job signs, business cards, direct mail, yellow pages, newspaper ads, neighbor's referral, etc. to get around to making that phone call. If you're on top of your business, you should know how much lead-time it takes to get your phone to ring. At my company we know that when we want our phone to ring, we need to allow approximately eight days for our advertising to reach the potential customer before we'll get that call. Whatever your time frame is from original advertising exposure to the customer's call must now be added to the total time process from new lead to the job completion. Let's say that the average time it takes to get the phone to ring at our company is two weeks (rounded up from eight days). When we add that to the ten weeks we've already established for the job above, that puts our total job time at twelve weeks. That's a minimum of three months working time to get the money in hand to pay our bills.

I once conducted a study on the number of leads that a salesperson should take to achieve the most profitable sales-to-lead ratio. At the time I did the study, I was working for a large remodeling company with 36 full-time salespeople. The study took six months to complete. We found, after analyzing the data from over 4000 leads, that the best number of sales leads for a remodeling salesperson to take was 18 per month. That was the number that ended with the most profitable sales and highest dollar volume sold each month. The second best number of leads was 19 and the third best number was 16.

Now, assuming you've read all the chapters in this book up to this point, this is the information you now have about generating cash flow:

- You have the ability to get the phone to ring.
- You know how many calls you need to generate the leads required to get the total sales you need.
- You have a sales plan in place, and you know how to write a good contract.
- You know how much lead time it takes to generate the money you need to pay your bills.

Sure this sounds simple enough, but probably one third of those reading this book have never even thought of running the numbers out like this for their company. If every builder would do this exercise, the failure rate in this industry would be cut in half.

Operating Capital Reserve Account

Let's go back to our sample remodeling company. Even though we know how to generate the numbers we need and make the sales, what happens if things go into a slump for our company? Maybe the economy is bad and construction in general is off. We know, without even having to think about it, what that'll mean to our company three months from now. How will we be able to pay the bills? How do we keep our company afloat during those off times? The answer is a cash reserve account — money in the bank.

However, if sales aren't good and we're starting to get behind on our bills or taxes, how can we build a cash reserve from which to draw? This is a tough problem to solve, but we can do it. I'll tell you how.

First, let's assume that we've done the "Quarterly Overhead Review Form" that's at the end of Chapter 2. We know what our overhead numbers are and we know what kind of profit we want to make. We also know that anytime we don't make enough money on a given job, our profit goes to paying overhead bills first.

We haven't set up an O.C.R.A. yet, so we must figure out a way to guarantee that we'll be able to pay our bills until we can get one built up and no longer need outside help. The quickest and easiest way to do this is get a line of credit at our neighborhood bank or S & L, or through another lender.

Chapter 7: The Mathematics of Your Business

That way we'll be able to pay the bills and not have to worry about where the money is coming from when a cash crunch hits.

Setting Up an Operating Capital Reserve Account for a Remodeling Company

Let's assume that we get a line of credit at the bank. Once it's in place, we must begin working on developing an O.C.R.A. And, we must have the discipline to stick with our yearly budget if we're going to make this process work.

After much research, I've determined that an O.C.R.A. should be equivalent to four months of total overhead expenses for your company. Your banker will probably tell you that you need five years, your accountant will tell you that 24 months is appropriate. Let's stick with my figure of four months for our example. It works for my company. You and your accountant and banker can argue and adjust your capital reserve account as you see fit.

We've established the monthly overhead expense at our company at $3,125 and four times that amount is $12,500. That's what we'll need in our O.C.R.A. to see us through any short-term problems that we might have. That O.C.R.A. account should be treated just like a new overhead expense. We plug it into the overall formula for our company and then put a plan into action to make it happen. Obviously, this will change our necessary total sales for the coming year, and it will also give us some additional profit for our company. That profit may be hard to achieve, but it will be more profit nonetheless. Let's see how our numbers change.

First, short of an inheritance or some other windfall that we can plug into an O.C.R.A., we'll probably need to project this new expense over a period of time. I recommend that you build your first reserve account over a period of at least one year. If you're doing a low volume of business, as in our example, then you might want to extend it over a period of two years.

Let's assume that we're going to build this account over a two-year period. We'll need to add half of the $12,500, or $6,250, to our projected overhead expenses for each of the next two years. Now we have a new projected overhead expense of $43,750 ($37,500 + $6,250 = $43,750). We'll use the formula for increasing sales volume to assimilate an overhead expense (see Chapter 3, page 77) to determine the new sales volume we'll need to cover the cost of this new overhead item. We divide our new overhead expense by our present overhead percentage of 25 percent to get the following numbers:

$$\$43,750 \div .25 = \$175,000$$

We must sell an additional $25,000 next year to get one-half of our O.C.R.A. in place. That means that we must take 3.57 percent ($6,250 ÷ $175,000 = 0.0357) right off the top of each and every check that comes into our company and put it into our O.C.R.A. and no place else. It doesn't pass GO, it doesn't collect $200, and it doesn't go to buy some new "toy" that we think our crew just can't do without any longer. (I don't recommend that you ever budget more than 4 percent of your total sales collected for your reserve account.) Our markup will remain the same, but we must make those sales happen if we're going to stay in business.

What about the other types of construction companies? How do they approach this issue? Specialty contractors should take the same approach as the remodeling company in our example. New home construction contractors normally do far fewer jobs than remodeling or specialty contractors, so they'll have a little bit different approach to all this. Let's see how it works for them.

Setting Up an Operating Capital Reserve Account for a New Home Construction Company

The majority of contractors in the new home market do less than a million dollars in total volume each year. I'm concerned with those who build from two to a dozen homes a year, not the big guys who build tracts of 500 to 1,000 homes each year.

A new home construction contractor normally has a much longer lead-time from the first day of advertising to the day he hands the buyers the keys to their new home. This process can easily take six months to a year or more. If you're a new home contractor, you can develop your own time schedule for getting homes sold, built and collected following the outline for the remodeling contractor we just discussed. Again, let me emphasize that building new homes is a different business than doing remodeling work.

So let's assume that you have your time schedule in place, and now you need to determine what your O.C.R.A. should be. Since your lead-time for your jobs is longer, I recommend that your reserve account be equivalent to at least six months of your projected overhead expenses.

We'll use the numbers from our new home construction example in Chapter 3 to determine what your O.C.R.A. should be. We also need to determine what percentage of your total sales the O.C.R.A. should be, and the annual sales volume required to cover this additional overhead expense.

If you've been working along with me as I've developed the solutions to these problems, you should be able to work this problem before we review it. Why not get out your calculator and see if you can come up with the same answers that I do? Build your O.C.R.A. over an 18-month period. See if you can do it — you'll never know unless you give it a try.

Here's what we know about our sample company:

▌ First year job costs are $506,000

- Overhead expenses are $87,950
- Profit is $55,000
- Total dollar volume is $648,950
- Markup is 1.29

First we need to divide our overhead expense for the year in half to get our overhead for six months and the amount that we want for our new O.C.R.A.

$$\$87,950 \div 2 = \$43,975$$

Again, we're going to project this expense over the next 18 months rather than the two years that we used for the remodeling business.

1. Start by dividing our desired O.C.R.A. by 18 months to get the additional expense per month

   $$\$43,975 \div 18 = \$2,443$$

2. Multiply $2,443 by 12 and add the result to our current projected overhead expense of $87,950 to get our new total overhead expense for the first year.

   $$\$2,443 \times 12 = \$29,316$$
   $$\$29,316 + \$87,950 = \$117,266$$

3. Our previous overhead percentage was 13.55 percent ($87,950 ÷ $648,950 = 0.1355). To arrive at the total sales necessary to cover this new overhead expense for the next 12 months, divide $117,266 by 13.55 percent.

   $$\$117,266 \div 0.1355 = \$865,432$$

The difference between our old and new projected sales is $216,482 ($865,432 − $648,950 = $216,482). That means we'll have to build one or two more homes this year (depending on the average sales price of the homes we build) to get our O.C.R.A. up and working.

Our math tells us that we must take 3.39 percent ($29,316 ÷ $865,432 = 0.03387) right off the top of each check for the next 12 months to start building our new O.C.R.A. At the end of the 12 months, we need to re-evaluate where we are with the account and how much more we need to meet our goal of $43,975. We can then calculate the balance to be added during the next six months using the same method we just completed.

If our calculations had shown the percentage we need in new home construction to be over the 4 percent figure that I said you shouldn't go over, then we would have to refigure this problem and spread it out over 24 months instead of 18 months. I used these numbers to show you that this formula will work over any projected length of time, not just even years.

If you're in the commercial construction market, you'll have to determine which of the two methods we've used in our examples will work best in calculating the total you need for your O.C.R.A. One or the other method will work for you; it all depends on the volume of work you do and the time span you need to complete it.

Make It Happen

In the years that I've been in business, I've only known a few contractors that had an O.C.R.A. in place, and every one of those contractors has done very well. A couple of them have retired now, and they're financially secure and doing what they always hoped they'd be able to do when they retired. Their O.C.R.A. provided them with a financial base from which they could make good decisions, and they were never caught in that crunch of not having the money to pay bills.

A cash reserve will also give you added bargaining power with your lenders. You've heard the old saying that lenders will only lend you money if you can prove you don't need it? Well, it's true. I've been on both sides of that fence, and I prefer the side with the bargaining power. It makes a world of difference in how and what you can do in your business.

Creating this account is one of those decisions that you can easily put off, but it's a decision that you need to make now. Understanding your cash flow and how to use an Operating Capital Reserve Account are necessities in this business.

Margins vs. Markup

While I was gathering information for this book, I heard several comments about the number of contractors who use margins rather than markup. It seems to be a general trend, not related to any specific trade. I think using margins instead of markup is fine, as long as you know the difference between them.

Gut check time again! Stop right here, close the book, and write down your definition of margin. Don't be shy, give it a try. If you're going to use it, then you should know what it is. You should know the subject inside and out. If you can write out a definition that's clear to anyone who reads it (anyone outside your company, that is), then you can probably use margins and make them work for you. If anyone who reads your definition isn't clear about what you mean, then I suggest you stick to using markup.

Again, this is one of those exercises that most people won't bother to do. They'll plow ahead assuming that their concept of margins is the same as the next person's. Those who want to be successful in this business will take the time to find out if their definition of margin is correct, and if not, revise it. Success is built on small steps. To reach the level of success that you want in this business, you need to take the steps.

Here are the definitions:

- Margin is the difference between what you sell a job for and what it costs you to build. Put another way, it's the total amount of your overhead and profit.

- Markup is your job costs times a number that allows you to sell a job at a given price that covers your job costs and overhead and also provides you with a profit.

When you look at these definitions, you can see that using a margin takes more mathematics to figure out than using a single number, as you do with markup. That's why I prefer markup. It's easier. We're going to take a look at using margins anyway, just to set your mind at ease. Using margins can be very misleading and get the unwary into all kinds of trouble.

First, let's review the terms that we'll be looking at.

- Markup is a number that you multiply your job costs by to give you a sales price that covers job costs and overhead expenses and provides you with a net profit for each job.

- Gross margin refers to the combination of overhead and profit for any given sale. A 30 percent margin on a $100,000 sale gives you $30,000 to cover overhead and leave you with a profit.

- Net margin refers to profit alone. So if you were to consider an 8 percent net profit satisfactory, then on that same $100,000 sale, you'd expect to make $8,000.

So far so good. Your margin can be the difference between your sales price and your job costs, or your margin can be your overhead and profit. Let's use the same numbers we used earlier in the book to arrive at markup to calculate your margin, and see how they compare.

1. ***Dollar Volume Built − Job Cost = Margin***

    $150,000 − $100,500 = $49,500

    $49,500 is our margin for the year.

Then, to find our margin percentage:

2. ***Margin ÷ Volume Built = Margin Percentage***

    $49,500 ÷ $150,000 = 33 percent

If a contractor tells you that he's trying to make a 33 percent margin on his work, you know he's using a 1.49 markup. In theory, you'll always have the same percentage of margin with a particular markup.

Markup & Profit: A Contractor's Guide **179**

Problems Using Margin

Let's review a couple of problems that frequently come up when contractors with a less-than-thorough understanding of margin try to use this approach to arrive at a sales price for their work.

The problem that I hear most often occurs when a contractor starts thinking that a 30 percent margin and a 1.30 markup are one and the same. They're not. A margin of 30 percent for a sale of $100,000 is $30,000. That leaves $70,000 to build the job, right? Let's take that same job cost of $70,000 and hit it with a 1.30 markup (1.30 × $70,000). That gives us a sales price of $91,000. That's $9,000 less. It takes a 1.43 markup to get a sales price of $100,000. Back to the drawing board. You can see how a contractor could end up if he uses these methods as though they were interchangeable.

Then there's the contractor who starts talking about the "big margins" he makes. He neglects to tell you about the small dollars he's working with. He thinks that a 30 percent margin is good. If you have annual sales of $975,000, a 30 percent margin is a fair number. But what if he's only doing $175,000 in sales for the year? With a 30 percent margin, he'll only have about $52,500 to pay all his bills (including his own wage that he needs to feed his family) and make a profit for the year. Again, you must remember that profit isn't something the owner stuffs in his pocket. Profit belongs to the company. So $52,000 isn't a good number. Something or someone is going to go wanting.

This second scenario is not unique to the construction industry. This misconception occurs in many businesses when owners start getting percentages and margins running through their heads instead of dealing with actual numbers.

Now, some of you may have just discovered "margins" and think that this is the more sophisticated or intelligent way to go. While it may be "politically correct" to follow this trend, and you may indeed turn heads, the reality is that using margins can get you into as much trouble as using someone else's markup. There's no correct margin for everyone, just as there's no correct markup for everyone. You have to figure your margin carefully and not be tempted to use the margin that everyone else uses. The most successful contractors are the ones that stick to the K.I.S. method of doing business. Don't get all hung up on dealing with percentages. Deal with actual numbers and you'll find that your business life is much simpler.

Break Even

Let's take a look at break-even points, how they're generated and how they can help you stay focused and make more money in your business. A break-even point is that point or time of year when your total income from all jobs sold, built and collected equals your total expenses for the year. In theory, until you reach your break-even point, you haven't made any money. That's the bad news. The good news is that from your break-even point on, you *are* making money. That's assuming, of course, that you follow through on what you should do as far as getting sales, using the proper markup and making collections.

Your break-even point is very important because you can see how you're doing in relation to your goals or projections for the year. Obviously, the earlier in the year that you reach your break-even point, the better your company is doing and the more money you'll make at the end of the year.

By the same token, if your company sales, production and collections aren't up to what they should be, your projected break-even point will be off, and you'll be able to clearly see just how far off you are. You can look at your break-even point as a goal, and of course, the name of the game is to exceed your goal.

The break-even point is normally calculated on the total projected sales for a company each year. Calculate what your break-even point will be right after you set your company goals each year. It's a fairly simple process requiring only basic math. There's a sample Break-Even Worksheet at the end of the book. Let's review a simple break-even point problem, and then follow that up with the break-even point for your company.

Figuring Your Break-Even Point

Let's use the figures from our sample remodeling company to calculate our break-even point for the year using the formula:

$$\text{Break Even} = \text{Total Overhead} \div \frac{(\text{Total Sales} - \text{Job Costs})}{\text{Total Sales}}$$

Here are the numbers we need to find the break-even point for a year and then for an average month:

- Total sales = $150,000
- Total overhead = $37,500
- Total job costs = $100,500

$$\text{Break even} = \$37{,}500 \div \frac{(\$150{,}000 - \$100{,}500)}{\$150{,}000}$$

$$\text{Break even} = \$37{,}500 \div \frac{\$49{,}500}{\$150{,}000}$$

$$\text{Break even} = \$37{,}500 \div 0.33$$

$$\text{Break even} = \$113{,}636$$

Our break-even point for the year is $113,636.

To find the break-even point for our average month, divide the break-even point for the year by 12:

$113,636 ÷ 12 = $9,470

To compute the approximate date that we'll reach our break-even point during the year, we divide our projected sales volume for the year by 12 to get our projected sales volume for each month. Then we divide $113,636 by our projected sales volume for the month.

$150,000 ÷ 12 = $12,500

$113,636 ÷ $12,500 = 9.09 or approximately September 3rd

As September is the ninth month of the year, our break-even point will fall on about the third day of September (30 days × 0.09 = 2.7 or 3 days). Easy enough!

Figures 7-1 and 7-2 show what a break-even chart might look like for a remodeling company and a new home construction company. A chart for a specialty contractor will probably look like the one for a remodeling company. Depending on the volume sold, built and collected, the contractor doing commercial construction, either new or remodeling, will probably have a chart that looks like the one for new home construction.

These charts are not based on the same dollar volume of work sold, built and collected. Nor are they designed to reflect a dollar volume for any company in particular. They are simply examples to show the value of a

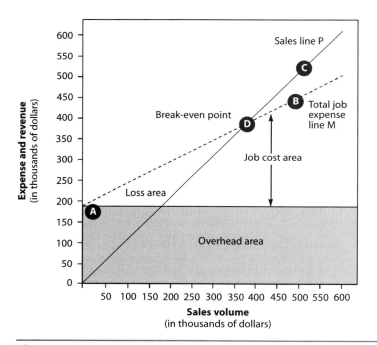

Figure 7-1
Break-even chart for a residential remodeling company

Chapter 7: The Mathematics of Your Business

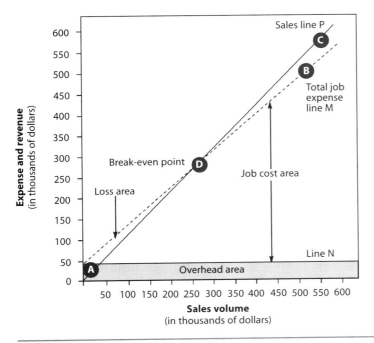

Figure 7-2
Break-even chart for a new home construction company

pictorial view of the cash flow through your company and how you can show your effective target date for your break-even point for the year. There is a blank Break-Even Point Worksheet in the back of the book that you can use for your own calculations.

Math, Formulas and Ratios for Your Company

Many of the things that we'll be looking at in this section are time-proven ideas that have been used in the construction business for decades. Some of them have been around since my grandfather served his carpentry apprenticeship early in this century, or even longer. Others are fairly new, but still important for the success of your business. Let's take a look at some ideas and formulas that might help with your cash flow problems.

Travel Time

We'll start with a formula for computing travel time for the occasional jobs that you take outside the normal work area of your company. I've used this formula for years, and though it isn't perfect, it's better than not having anything in your estimate to cover extended mileage and travel time to outlying jobs. The formula is:

Travel Time = $2/Person/Mile/Day

That means that you charge your customer $2 per employee per mile every day that your employees must travel to and from a job site in excess of their normal commuting range. Suppose, for example, that you've taken a job that's 30 miles from your office. Part of your employee job description might be a requirement that the employee be willing to drive to up to 20 miles from your office to a job site at no charge to the company. This distance will vary from company to company, but let's assume 20 miles for our example. Let's also assume that you'll have two people on the job a total of six days. To figure out the dollar amount that you should add to your estimate to cover travel time for your crew, plug your numbers into the formula.

Travel time = $2/person/mile/day
Travel time = $2 × 2 (employees) × 10 (miles) × 6 (days)
Travel time = $2 × 2 × 10 × 6 = $240

Will this formula work across the board for all companies? Probably not. However, it will work for most companies. The goal is to have a formula that you can use to fairly compensate your employees for the extra miles they must drive, without running the cost of the job up so high that you eliminate your company from the selection process. Start with this formula and adjust it as needed. You should be able to have a formula worked out that suits your company and your employees in three to six months, depending on the number of outlying jobs you take.

One way my company controls the problem of long distance traveling is to simply not advertise in areas that are further away from the office than we expect our employees to travel. That way we're not tempted to overextend ourselves. There isn't a perfect circle of areas to work in around the office, but we pretty well know which communities we want to work in. Since the large majority of our advertising is direct mail, we simply select the Zip codes we want and mail only to those areas.

Another thing you must watch for in paying travel time is consistency. Employees are funny. If they expect to get paid for something, you must pay them. They'll get mad and leave your company over a few dollars, even though they've worked for you for years. I know a contractor who refused to pay his employees for four days travel time to a job that was 18 miles out of his operating limit. He forgot to include the cost in his estimate, so he didn't think he should have to pay his employees if he wasn't getting paid. He lost two very good carpenters over this. Can you imagine? The contractor couldn't believe that his guys would quit over $144. He thought they were really dumb. But when he asked what I thought, I told the contractor that I thought he was the foolish one. It would cost him twice the $288 that he refused to pay to advertise and hire two new carpenters, not to mention job delays, and the cost to his reputation. Word gets around in the trades about who does and doesn't deal fairly with employees. No one wants to work for someone who'll pull a stunt like that!

Whenever you make an agreement with an employee, you must keep it. It doesn't make any difference what reason you have for wanting to back out; if you made the agreement, you stick to it. *And, when you owe — you pay!* You can't penalize an employee for your own incompetence.

Renting vs. Buying

What's the best approach when it comes to spending company capital on tools and equipment? Is it better to buy or to rent? Here's a tip for those of you who spend money now, and then wonder later why you did it. When determining whether it's better to buy an item or rent it, check your records and see how many times in the last year you've needed the particular tool or piece of equipment. Then:

- If you use an item **24 days** a year or less, you should continue to rent it.

- If you use an item **25 days** a year or more, you should buy it (providing you know you'll have a continued need for the item).

You can check the value of this advice. Go into the area where you keep your tools and equipment. If an item has dust on it, you probably should have rented it instead of buying it. If you're using a tool or piece of equipment more than 25 times a year, chances are dust won't collect on it between uses. There are exceptions of course, such as seasonal equipment like heaters or pipe thawing machines.

You probably already have most of the basic tools you need for the work you do. If you're not going to work on jobs yourself, but either hire employees, or subcontract the work out, make sure the people who're doing the work have their own tools. If they don't have their own tools, don't hire them. Why should your company take on the expense for someone else's tools of the trade?

The Value of Owner's Time vs. Hiring It Done

We covered this a bit earlier, but it still needs to be included here to ensure that company owners who're reading this book put the proper value on their time. It's easy to let this get by you. If you put a definite value on your time, it'll keep you from using your time ineffectively.

Let's assume that you set the value of your time to your company at $50 an hour. Now, suppose there's a job that needs to be done, one that you can do yourself. If you can pay someone else less than $50 an hour to do the work, that's what you should do. If the cost of that particular job runs more than $50 an hour, then you can decide if you can afford to take the time away from your other duties to do the job. If you already have the money in the job to hire that function done, don't get involved. If you don't have the money allotted, and you do have the time, you can do it and charge the company for your wages.

Now, if the job has something to do with the running or promotion of your business, that's another matter. Do you have the knowledge and ability to do the job effectively, or could someone else do it better? What would be best for your company? If you can make more money for your company by hiring a professional while you work on something else, then you should do that. If your time is best spent on this project and it'll put money in your bank account, then you do it.

Number Stuff

It's time to talk a little about accounting tools that will help you analyze where you're going with your business. Your CPA has probably given you some of this advice already. You have to decide how involved you want to get in the accounting and bookkeeping end of your business. You can usually hire someone to prepare reports and lay them on your desk far cheaper than you can do them yourself. However, you still need to know how to measure the financial health of your company.

Net Profit

You should be making a minimum of 8 percent net profit. Remember our earlier definition of profit? It's what's left after *all* the bills are paid.

Do you think that making an 8 percent net profit is too much? It might interest you to know that those designer running shoes that your kid *had* to have and you paid $95 for cost somewhere between $6 and $10 to make. What does that say about the profit built into that sale? Double the manufacturing cost of the shoes for shipping, display, advertising, and so on, and the company selling the shoe is still making a 210 percent profit! The markup on furniture is generally 100 percent or more. Doctors and lawyers all shoot for a minimum of 20 to 25 percent profit, and the smarter ones push that to 50 percent. So what's wrong with you getting 8 percent profit? You're worth it!

Gross Profit Ratio

As you read through this information on ratios, please keep in mind that some accountants and accounting references may refer to these using different terms. If I've called a ratio or a process by a name other than the one you're familiar with, please bear with me. It's an understanding of the process you need, more than the name of the process.

The gross profit ratio (also know as gross profit margin) is the relationship between your gross profit (the difference between the sales price and the job costs of a given job) and your total sales. To find it, you add up the gross profit on all your jobs in a given period and divide that by your total dollar volume sold for that period. That'll give you a percentage figure. You can do this per job, monthly, quarterly, or yearly. Obviously, the higher your percentage, the better you've done. This ratio gives you the ability to measure how well you've done this month compared to last month, or this period compared to the same period a year ago. It'll help you spot problem areas such as an incorrect sales price on work or production or scheduling problems.

Here's an example of how you figure your gross profit ratio on a single job:

Sales price of job	$85,659
Job costs	−64,210
Gross profit	$21,449
Gross Profit Ratio	$\dfrac{\$21,449}{\$85,659}$ or 25.04%

This ratio also allows you to see the end result of your markup, and points out the discrepancy if you've ignored the numbers and used a markup other than what the math dictates for your company. And, if you're one of those who insist on using a sliding scale markup, this ratio will give you the ability to check your total jobs each month or quarter to be sure that you're maintaining the proper profit percentage overall.

When you begin using this ratio, your percent will vary. What you want to do is establish a norm for your company, and then work on improving that with each job or in each period that you're measuring. Here are some suggestions to start with. New home construction should be in the range of 20 to 30 percent. You'd better have a well-thought-out reason and good justification for being under 20 percent. Remodeling and specialty contractors should probably be in the 25 to 40 percent range. Again, if you're under 25 percent, you'd better have a very good reason for it. Those doing commercial work may well be down in the 10 to 25 percent range, depending on the volume they do and how efficient they are. A land developer would fall at the other end of the spectrum. Because of the risk factor and the length of time that they're usually involved with any given project, their ratio is in the 40 to 60 percent range.

Gross Profit Per Manhour

I recently talked to a remodeling contractor who calculates his gross profit per manhour to measure the production of his staff. If you do your bookkeeping on computer, you could easily track this and it would be a good measuring stick for your company. However, you must set some standards on how to apply this measurement, and then stick to those standards. An evaluation of your production will not work if it's not consistently applied.

Current Ratio

The current ratio measures the liquidity or solvency of your company. In other words, it's a measurement of your ability to pay your bills with your available money.

The formula for the ratio is:

$$\frac{Current\ Assets}{Current\ Liabilities} = Working\ Capital$$

The lower your debt, the higher your asset to liability ratio will be. Ideally, you would like this ratio to be 2/1 or just 2. That would mean that you have twice as much money in the bank (current assets) than you have bills to pay (liabilities). In all probability, however, it may be as low as 1.5 for those of us in construction. A typical remodeling company's figures might look something like this:

$$\frac{\$48{,}692}{\$29{,}457} = 1.65$$

Again, this is a formula that you must use for a while in order to establish a norm for your company, and then you can work on improving it.

Cost-of-Sales Ratio

You've probably been using the cost-of-sales ratio and the gross profit ratio for years without knowing the fancy names for them. The cost-of-sales ratio is your total cost of sales (your job costs and your overhead expenses) all lumped into one number and divided by your total sales for the period.

$$Cost\text{-}of\text{-}Sales\ Ratio = \frac{Cost\ of\ Sales}{Sales}$$

The ratio is expressed as a percentage. If you're like 80 to 90 percent of the contractors who'll be reading this book, your sales will probably not exceed $1.5 to $2 million dollars of work per year. In that case, your cost-of-sales ratio should never exceed 92 percent. In other words, your costs and your overhead can't exceed 92 percent of your sales price if you're going to make a minimum 8 percent net profit!

Marketing Expense Ratio

Most of us have talked about the marketing expense ratio before, but probably not given much thought to the real name for it. It tells you what percent of your total budget for a given period is being spent for advertising and marketing.

$$Marketing\ Expense\ Ratio = \frac{Marketing\ Expenses}{Sales}$$

The percentage for new home construction will normally run 2 to 5 percent, remodeling 3 to 5 percent and specialty contracting 1 to 2 percent. Keep in mind, however, that if your numbers are different from these and you're making the minimum 8 percent net profit, you don't need to change. These numbers are considered the norms in the industry. They are just a place to start; to give you an idea of what others in the industry are doing.

There are a number of other ratios that are used by accountants to describe various aspects of business. If you want to know more about them, your accountant would probably be willing to discuss them with you.

Chart of Accounts

A chart of accounts is a listing of all your company's major expense items, with each item given its own account number. Every expense your company has will fall under one of these accounts. It's used primarily by your bookkeeper and accountant to track your expenses. Figure 7-3 shows a typical chart of accounts for a construction company. I think every contractor I've ever met has a different chart of accounts. New home construction contractors have one kind, remodelers another, specialty contractors another, and so on. Can you imagine the poor accountant that does the books for several different companies, each involved in a different type of construction work?

Find a good chart of accounts that has been prepared by one of the major construction associations and use it. Your accountant will probably love you for it. Or, use the chart of accounts that your accountant recommends. Don't spend your time inventing something that others have spent years perfecting.

Computers

The computer age is finally gaining ground in the construction industry. Contractors are slow to change methods of doing almost everything. A recent survey done by a building association on the West Coast showed that only about 30 percent of their members were doing their estimating using a computerized estimating program. The other 70 percent were still using paper and pencil. However, a computer, properly employed, will reduce the number of hours you have to spend on repetitious paperwork tasks, as well as reduce the number of errors in your accounting, estimating, and contract writing.

Contractors who incorporate the use of computers in their day-to-day activities know where their companies stand financially on a daily basis and can make sound money decisions. A few years ago, I wrote an article for *Western Remodeling Magazine* about the cost of not automating a remodeling company. I based the article on a study I had done during the last half

Chapter 7: The Mathematics of Your Business

Figure 7-3
Chart of accounts

Harbour Towne Construction
Chart of Accounts

Account ID	Account Description	Active	Account Type
10000	Petty Cash	Yes	Cash
10200	Regular Checking Account	Yes	Cash
10400	Savings Account	Yes	Cash
10600	Investments - Money Market	Yes	Cash
10700	Investments - Certificate of Deposit	Yes	Cash
11000	Accounts Receivable	Yes	Accounts Receivable
11400	Other Receivables	Yes	Accounts Receivable
11500	Allowance for Bad Debt	Yes	Accounts Receivable
11800	Unearned Income	Yes	Accounts Receivable
12000	Inventory	Yes	Inventory
14000	Prepaid Expenses	Yes	Other Current Assets
14100	Employee Advances	Yes	Other Current Assets
14200	Notes Receivable - Current	Yes	Other Current Assets
14250	Pre-Billing	Yes	Other Current Assets
14300	Work in Progress - Backlog	Yes	Other Current Assets
14400	Materials - Unused & Unassigned	Yes	Other Current Assets
14500	Deposits	Yes	Other Current Assets
14700	Other Current Assets	Yes	Other Current Assets
15000	Furniture and Fixtures	Yes	Fixed Assets
15100	Office Equipment	Yes	Fixed Assets
15200	Automobiles	Yes	Fixed Assets
15300	Machinery & Equipment	Yes	Fixed Assets
15400	Leasehold Improvements	Yes	Fixed Assets
17000	Accumulated Depreciation	Yes	Accumulated Depreciation
19200	Loans to Officers	Yes	Other Assets
19300	Loans to Shareholders	Yes	Other Assets
20000	Accounts Payable	Yes	Accounts Payable
20050	Commissions Payable	Yes	Other Current Liabilities
20100	Retainage Payable	Yes	Other Current Liabilities
20200	Workers' Comp Payable	Yes	Other Current Liabilities
21000	401K Deduction	Yes	Other Current Liabilities
21100	Accrued Expenses - Other	Yes	Other Current Liabilities
21300	Group Health Insurance Payable	Yes	Other Current Liabilities
22000	Notes Payable - Short Term	Yes	Other Current Liabilities
23000	Federal Payroll W/H Tax - Payable	Yes	Other Current Liabilities
23100	State Payroll W/H Tax - Payable	Yes	Other Current Liabilities
23200	FICA Taxes - Payable	Yes	Other Current Liabilities
23250	Medicare Taxes - Payable	Yes	Other Current Liabilities
23300	FUTA Taxes - Payable	Yes	Other Current Liabilities
23400	SUTA Taxes - Payable	Yes	Other Current Liabilities
23450	Other PR Withholdings	Yes	Other Current Liabilities
23500	Sales Tax Payable	Yes	Other Current Liabilities
24700	Other Current Liabilities	Yes	Other Current Liabilities
24800	Suspense - Clearing Account	Yes	Other Current Liabilities
27000	Notes Payable	Yes	Long Term Liabilities

Figure 7-3
Chart of accounts (continued)

Account ID	Account Description	Active	Account Type
27100	Office Furniture Payable	Yes	Long Term Liabilities
27200	Computer Equipment Payable	Yes	Long Term Liabilities
27300	Loan Payable TPD	Yes	Long Term Liabilities
27400	Credit Line Payable	Yes	Long Term Liabilities
28000	Capital Stock	Yes	Equity-doesn't close
28500	Added Paid-in Capital	Yes	Equity-doesn't close
28999	Beginning Balance Equity	Yes	Equity-doesn't close
29000	Retained Earnings	Yes	Equity-Retained Earnings
30000	Remodeling Revenue	Yes	Income
31000	Repair Revenue	Yes	Income
31500	Design Revenue	Yes	Income
32000	Miscellaneous Jobs Revenue	Yes	Income
32150	Discounts Given	Yes	Income
32250	Unearned Income	Yes	Income
32500	Discounts Earned	Yes	Income
33000	Interest Income	Yes	Income
33100	Commissions Earned	Yes	Income
33200	Non-Profit Revenue	Yes	Income
33250	Other Income	Yes	Income
40000	Labor Costs	Yes	Cost of Sales
42000	Material Costs	Yes	Cost of Sales
42900	Shipping Charges Reimbursed	Yes	Cost of Sales
43000	Sub Labor Costs	Yes	Cost of Sales
44000	Sub Material Costs	Yes	Cost of Sales
45000	Equipment Costs	Yes	Cost of Sales
46000	Other Costs	Yes	Cost of Sales
46500	Materials Non-Profit	Yes	Cost of Sales
47000	Sales Tax	Yes	Cost of Sales
48000	Insurance - General Liability	Yes	Cost of Sales
48100	Insurance - Workers' Comp	Yes	Cost of Sales
48200	Medical Expense - Field Employee	Yes	Cost of Sales
48300	Employee Benefits - Field	Yes	Cost of Sales
48400	Disability Insurance - Field	Yes	Cost of Sales
48500	Health Insurance - Field	Yes	Cost of Sales
48600	Life Insurance - Field	Yes	Cost of Sales
48700	Holiday Pay - Field	Yes	Cost of Sales
48800	Vacation Pay - Field	Yes	Cost of Sales
48850	Bonus - Field	Yes	Cost of Sales
48900	FICA Expense - Field	Yes	Cost of Sales
48925	Medicare Expense - Field	Yes	Cost of Sales
48950	SUTA Tax Expense - Field	Yes	Cost of Sales
48975	FUTA Tax Expense - Field	Yes	Cost of Sales
49000	Pagers	Yes	Cost of Sales
49050	Radios/Cell Phones	Yes	Cost of Sales
49100	Repairs & Maint - Tools/Equipment	Yes	Cost of Sales
49200	Supervision - Production Manager	Yes	Cost of Sales
49300	Supplies - Miscellaneous	Yes	Cost of Sales
49400	Tool Expense	Yes	Cost of Sales
49500	Tool Allowance	Yes	Cost of Sales

Figure 7-3
Chart of accounts (continued)

Account ID	Account Description	Active	Account Type
49600	Vehicle Expense - Field	Yes	Cost of Sales
49700	Warranty Expense	Yes	Cost of Sales
60000	Advertising Expense	Yes	Expenses
61000	Commissions	Yes	Expenses
70000	Salary - Officers	Yes	Expenses
70100	Salary - Office Staff	Yes	Expenses
70200	FICA Expense - Office	Yes	Expenses
70300	Medicare Expense - Office	Yes	Expenses
70325	SUTA Tax Expense - Office	Yes	Expenses
70350	FUTA Tax Expense - Office	Yes	Expenses
70400	Bonuses	Yes	Expenses
70500	Employee Benefits - Office	Yes	Expenses
70525	Disability Insurance - Office	Yes	Expenses
70550	Health Insurance - Office	Yes	Expenses
70575	Life Insurance - Office	Yes	Expenses
71100	Association Fees	Yes	Expenses
71200	Auto Expense - Office	Yes	Expenses
71250	Auto Expense Repairs & Maintenance	Yes	Expenses
71300	Bad Debt Expense	Yes	Expenses
71400	Bank Charges	Yes	Expenses
71500	Computer Software Support	Yes	Expenses
71600	Contributions	Yes	Expenses
71700	Depreciation	Yes	Expenses
71800	Dues and Subscriptions Expense	Yes	Expenses
71900	Education & Training	Yes	Expenses
72000	Entertainment/Meals	Yes	Expenses
72100	Insurance - Auto	Yes	Expenses
72200	Interest Expense	Yes	Expenses
72300	Licenses	Yes	Expenses
72500	Miscellaneous Expense	Yes	Expenses
72600	Office Equipment	Yes	Expenses
72650	Office Equipment Repair & Maintenance	Yes	Expenses
72700	Office Supplies	Yes	Expenses
72800	Postage Expense	Yes	Expenses
72900	Professional Development	Yes	Expenses
73100	Professional Fees - Accounting	Yes	Expenses
73200	Professional Fees - Legal	Yes	Expenses
73300	Rent - Storage	Yes	Expenses
73400	Rent - Office	Yes	Expenses
73500	Leasehold Improvement	Yes	Expenses
74000	Repair & Maintenance - Vehicles	Yes	Expenses
74100	Taxes - Other	Yes	Expenses
74200	Telephone	Yes	Expenses
74300	Travel	Yes	Expenses
74400	Truck Lease	Yes	Expenses
74500	Utilities	Yes	Expenses
81000	Taxes - Income Federal	Yes	Expenses
82000	Taxes - Income State	Yes	Expenses
90000	Gain/Loss on Sale of Assets	Yes	Expenses

of 1992. I found that contractors who had automated their companies, and were doing at least 75 percent of the tasks that could be done by computer using computer programs (estimating, contract writing, accounting, job costing, letter writing, lead tracking, etc.), were saving a considerable amount of money. It worked out to approximately $1,305 in costs for every $25,000 in work sold, built and collected. It doesn't take a rocket scientist to figure out that computers not only make life easier, but they can also reduce your overhead expenses. A good computer system for your company will pay for itself very quickly, and probably give you some "time off for good behavior!"

What to Buy

If you have a computer system already, you may find it's time to upgrade. In my company we upgrade at least once a year and change out our whole system every two years or so. That's how fast changes are occurring in computer technology. You may not want to put that much money into your system, but here are some of the basics that I think you need.

You should have a Pentium-class computer with at least a one-gigabyte hard drive. You also need a modem, CD-ROM drive and tape backup. I like to use at least a 15-inch color monitor, and I think a color printer is essential if you're going to produce your own forms with your company letterhead. Of course, these standards may be changed by the time this book comes out. Technology changes that fast! Don't cut corners when you buy your computer equipment. You get what you pay for.

Software

Most computers today come with a preloaded software package. They usually include a word-processing program, a spreadsheet, day planner of some kind, some games and a program that will provide you with Internet access. There may be other options as well, depending on what the supplier thinks is relevant. Whatever else you get in software is your personal choice. Pick and choose those programs that you can relate to and learn to use easily. No two people will look at software exactly the same way, so don't expect everything that comes on your computer to be of use to you. However, a good general office suite of programs is pretty hard to beat as far as the number of business-related jobs you can do.

What to buy? Here are some suggestions for a few programs you need to have, and beyond these it's a matter of choosing the software that you think will do you and your company the most good. In the back of the book you'll find a list of some software development companies that offer products that you can use successfully in the construction industry. Call or write them to find out what programs they have available, what the programs cost, and how you can incorporate them into the everyday business routine of your company.

Lead Tracking — The first program you need is one that will track your leads. You must be able to keep track of the contacts you've made, and work them on a long-term basis. You can get a good program in the $75 to $150 price range.

Estimating — The second program you should buy is an estimating program. In the fast-paced world we work in, a successful contractor can't afford to take two or three weeks to get an estimate back to their customers. You must be able to get back to them within two or three days, and with accurate numbers. There are several estimating programs available from the companies listed in the back of the book. They range in price from about $40 to $500. A good estimating program is essential to your company.

Proposal/Contract Writing — The next program you need is one that will help you with proposal and contract writing. The computer never forgets the contract language that you need, and these programs will provide you with a contract that will make a good and accurate presentation for your customers. Be very careful here, however. Don't get a program that simply downloads the line items from an estimating program, then calls it a proposal or contract. In most cases, if your customers could read your estimate sheet, you wouldn't be necessary. Get a program with good, all encompassing contract language. These run between $75 to $250.

Accounting — A good accounting program is an indispensable asset to your business. Intuit has a couple of good programs that work well for the construction industry. They are QuickBooks and QuickBooks Pro. They run about $200, more or less, depending on who you buy them from. Other programs are available for $150 to $500. Again, check the list of companies in the back of the book.

Job Costing — You need to get a job-costing program that's simple to use and gives you pertinent information. If your accounting program has a job-cost module, then you won't need a separate program. If you do need one, make sure it's easy to use and understand. You don't want something that only your accountant can use. About $75 to $250 will get you a very good program.

CAD Programs — Oh, the joy of making your own drawings! These programs can be a blessing or an absolute nightmare. Think carefully about purchasing one these programs. They can be a big investment. Do your homework. Some are easier to use than others, but none could be classified as turn-it-on-and-use-it type software in spite of what some vendors will tell you. You need to decide if it's more cost effective to do your own drawings or pay someone else to do what they do best. Most contractors I know who bought CAD programs never used them, and the programs ended up collecting dust on a shelf. I don't know how many times I've heard "I just don't have the time to learn how to use the darned thing!" The programs were too difficult and too time-consuming for them to learn, or they simply weren't cost effective.

Some Advice on Software

Even though you're the boss, don't try to jam your choice of programs down the throats of your staff. Maybe you like a particular program, but that doesn't mean that it'll be easy for everyone else in your company to learn. We all look at work a little differently. This applies to accounting packages in particular. Let software buying be a group decision. If you try and force your staff to use a program they don't like, they simply won't use it.

Stay away from "piggyback" programs. These are templates that have been written for such programs as Excel, Lotus and other parent programs. If you buy one of these piggyback programs, you're often paying for two complete programs and getting only the parent program and a template. It's cheaper to buy a program written for the specific job that you want done. Not only will it work better, but you'll only have to learn one program instead of two.

"The second program you should buy is an estimating program. In the fast-paced world we work in, a successful contractor can't afford to take two or three weeks to get an estimate back to their customers."

I've read statements in various trade magazines and books that claim you can buy a good software spreadsheet program and use it to build your own database for estimating. I have a hard time believing that anyone would give credence to this nonsense. Spreadsheets are not designed to do estimating. Even if you have the ability to build a database for yourself, think of the time that it would take you to do it! This is another one of those times when you should let the experts do it for you — trust me, it's a lot less expensive, far less frustrating, and you'll end up with a better program. A good estimating program will have you producing estimates in about 6 to 8 hours — just the time you need to learn the program.

Sometimes I hear people say, "None of the estimating systems operate the way I want them to. I do things different!" All I can say in response is that sometimes you have to be a little bit flexible. If you can't adjust your thinking in order to run an estimating program (even though it may not be exactly what you want), then I suspect you're going to have trouble in other areas of your business as well. You have to learn to adjust to new ideas and new ways of doing things if you're going to keep up in this industry. The secret to good estimating is to do the same thing, the same way, every single time. Pick a program that you can work with, even if it isn't perfect. Focus your time and energy on getting your estimates done and your jobs built and collected rather than having everything your way.

Borrowing or lending software is something that I'm particularly sensitive about. There's no such thing as "borrowing" software. If you're not paying royalties to the developer of that software, it's stealing — and you're in violation of the purchase agreement. If you want a program, buy it. You don't supply two bathrooms for the price of one to your customers, do you? You don't give your design ideas away to your clients so they can pass them on to their friends, neighbors and family, do you? If you buy one package, you're entitled to use that program on one computer, not two or three or however many you have in your company. It's intellectual property, and the people who assembled the program should be paid for their efforts. Using borrowed programs is just plain dishonest.

Get a Fax Machine

A Fax machine, like a cellular telephone and a pager, is essential in today's business world. You have to be able to communicate with other people in the business world. Fax machines save so much time it's hard to put a price tag on their value to your business. They are marvelous machines.

Chapter 8

Bean Counters and Your Business

Bean counters aren't necessarily accountants or bookkeepers. Some work in banks, insurance companies, planning departments, inspection agencies, perhaps even in your own office. Now there's a sobering thought! They've been placed on earth to try to make you fit into a mold. In other words, to make your life in construction miserable. I have gray hair, and I'm sure it's entirely due to the well-meaning but insidious acts of bean counters. So how can you navigate this sea of sharks without getting bitten in your aspirations?

Let's look at some of the precautions that you can take to stop their well-meaning mischief and keep them on track helping you and your business rather than wasting your time.

Dealing With the Bean-Counter Mentality

First, you must understand the bean-counter mentality. They're wired just a little different from you and me. If we're wired with four or five circuits to keep us going, they're wired with twenty-seven. That's their nature. We try to keep things simple, as in K.I.S. They like to add a twist, a turn, an extra step or one more report just to clarify this or satisfy that. It goes back to some pagan primordial need to make things as complicated as possible. The secret to dealing with them is to make sure that you give them so much stuff to work with that you send them off on an emotional high. By the time they plow through everything, they'll be too tired to ask for more.

Use this tactic on the plan inspection department that wants "engineered drawings" for a simple room addition, the banker who wants the underground tanks inspected complete with soil tests (at your expense), and the accountant who wants to see your tax returns for the last 14 years.

Whenever you deal with anyone who has a bean-counter mentality, be sure you have all your paperwork in order, and you include everything they've asked for — and more. For example, when submitting plans to the planning department, have someone double-check your paperwork. Even when you're satisfied that everything for your project is complete and you've complied with all their requests, someone else may look through the documentation and see that there's something they think you should include. It may not necessarily be something you've forgotten, just an item that could possibly slow down your progress through the paperwork mill if you don't include it. Be sure your plans are neatly drawn, that they look very official, and have dates, names, addresses, and phone numbers. Don't forget the schedules, and lots of details. If it looks like you know what you're doing, chances are your project will fly right through the plan inspection process prior to your permit being issued for your job. Think it through, make your plan, and act accordingly.

When you go in to establish a line of credit prior to getting your Operating Capitol Reserve Account set up, you'll get a stack of papers from your lender to fill out. Make sure that you return with a pile of papers that's at least four times as thick as the one they gave you. An application for a line of credit requires a half-inch-thick pile, minimum. If your M.O.M. is complete, send a copy of that along with your application. Lenders love that stuff. If you have a business plan completed, send that along as well. Ninety-nine percent of loan applicants aren't this well prepared. Your lender will be in absolute ecstasy! How do you think that will affect your chances? It may take you a few minutes longer to organize everything, but it'll save you hours of work, time and effort down the road.

Bean Counters in Your Office

What about the bean-counter types you find in your office? Bean counters, by their very nature, are methodical and detail-oriented workers. If you understand how they work, you can use them to help you keep things neat, tidy and balanced in your business. Since most contractors don't have the ability to do this on their own, you'll end up with a win-win situation. Most problems that come up in your office will usually involve your bookkeeping system. Since most bookkeeping systems are far more complicated than necessary, you may end up in a situation where your bookkeeping staff understands more about your system than you do. Then, when you're called upon to make a bookkeeping decision, you're lost. Most contractors just tell their bookkeeper to take care of it, and go back to building their jobs — something they enjoy and that makes sense to them. But that's how contractors lose touch with their business. They can no longer tell you if they're making money or not.

You must establish a good system of communications between your bookkeeping staff and yourself, or whoever makes the financial decisions for your company. Once the bookkeeper understands what's needed, he or she can supply you with the information required to make day-to-day financial decisions and keep you in touch with your business. *You must work together*. When communication breaks down, then problems begin. Don't let someone else assume responsibility for decisions that should be yours.

Make sure your bookkeeping department has clear, well-though-out instructions about what you need and why you need it. Put a good plan into effect based on the advice of your CPA. Don't turn your back on this area of your business. It's critical that your bookkeeping process run smoothly and fit in with what your CPA needs, since your CPA also does your taxes and keeps track of your profits and losses.

A good accounting (bookkeeping) system is very important to your business. Take the time to review yours. It makes a difference in how you appear to your lender, accountant, financial advisor or anyone else that works with your business.

Financing with Two Bankers and One Broker

In our business it's important that we locate good financing for our customers. We don't necessarily have to get involved in the actual financing process, but we need to be sure that we have it on line. Over the years I've found that it's a good idea to have at least three separate sources of financing, one of which should be a mortgage broker.

Every area of the country does things just a little differently when it comes to financing. So one factor in determining what to look for in a lender will be your state laws, and another will be the regulations set by the lender. You can choose who you want to recommend to your clients. If a lender insists that all loans be placed in an escrow account, and it's a state law that you must do this for home improvements or new home construction, then that's something you'll just have to deal with. On the other hand, if the lender insists on this approach, and you have the option of using a lender that doesn't, then the choice is clear. The fewer people involved in handling the money coming from your customer's account to your account, the better.

One of the things that we do at our company when our remodeling customers set up financing is to make it very clear that we only deal with them (the customer) and not their lender on all financial matters pertaining to their job. We don't talk with lenders, period. If we have to talk to their lender for any reason, our rates are $150 per hour, paid to us *before* we talk to the lender. That may seem a bit hard-nosed, but that's OK with me. The last thing in the world I want to do is deal with some lender who thinks he can tell me how I should run my job, or who threatens to not release the next draw until this or that is corrected. Keep the job between you and your customer, and keep all others, including the lender, out of the loop.

If you build new homes, that's another matter. You have to play the game a little differently. A lot of lenders insist on progress inspections before they release draws against work completed on a given home or building. If you get crosswise with the inspector, it could mean a delay of a month or more before you get your money for the work you've done. Of course, there are two sides to this situation. If you do shoddy work, don't keep the job clean, and don't stick to the time schedule for the job, then you deserve to have your money tied up. You don't pay your subs for bad work, so why should a lender pay you for bad work? However, if you do good work but you find yourself getting wound up in nit-picky nonsense with the inspector, then you need to take another approach. You need to develop a relationship with someone at the lending institution who can stomp on this as soon as it starts so that you can get your draws approved on schedule. This is a touchy area, and trying to bull your way through it is seldom effective. I've tried a few times. Diplomacy is the best approach.

"Make an effort to establish a good working relationship with your lenders. It will serve you well."

Have a little heart-to-heart talk with someone at the lending institution. You do have some leverage there, contrary to what almost all lenders will tell you. You can take your future business elsewhere. Lenders are in business to lend money. They want your business. It's really quite simple. If dealing with a particular lender becomes difficult, tell them to ease up or you'll take your business to the competition. If your credit is good and your reputation is good, that's not just an idle threat. You can do it. You need to get your money on time so that you and your workers and your suppliers can be paid on time. You have obligations to meet. If you don't take steps to be sure you get your money on time, who's at fault? It's your business, so you have to be sure that you make things happen when they're supposed to. Make an effort to establish a good working relationship with your lenders. It will serve you well.

Choosing Your Bankers

While banks aren't known for having problems getting loans arranged and closing them on time, it does happen. My personal experience is that good service is generated by the individuals working in the loan department at the bank rather than by any particular bank policy. And problems arise from individuals as well. So, when you first start doing business with a bank, and you think there's a good chance that you'll want to work with them again, or recommend them to potential customers, then you need to sit down and talk with the loan officers.

This is the approach I recommend for introducing yourself to a loan officer:

1. Call and make an appointment. That will set the stage and let them know that you're not some flaky contractor that just walked in off the street on a whim.

2. Be on time for the appointment and dress appropriately. The CEO of a large company once told me that "bankers understand three-piece suits." That piece of information has been invaluable to me over the years. When you go to see a banker, dress like a banker. That puts you on even ground with them from the start.

3. Prepare a brief on you and your company that you can lay on the desk for the banker to see. That will let him or her know that you're a serious businessman, not someone looking for favors. It should include a company history, your resume, job pictures, the names and phone numbers of happy customers, and the appropriate financial papers.

You need to learn the bank's procedure for granting loans. Once you establish how they make their loans, you must let them know, politely, that when you refer a customer to them, they should never say anything to that customer about getting "three bids." You don't care if they do that with other contractors, but when your clients come through the door, the only thing they should discuss is:

- What they need from the customer to get the loan started
- How fast they can process the loan
- When the loan will close

Many lenders have a list of contractors that they deal with frequently and refer their clients to. Make sure that you're on that lender's list and that they will reciprocate your referral by referring potential clients to you. I only deal with lenders who keep my company on the list of "the three good contractors that we recommend."

Make sure that any lender you recommend will treat your clients fairly. Advise your clients to watch out for excess charges in the form of fees. Be sure that they know that any fees tacked onto a loan agreement should be explained to them before they make a commitment to that loan. Your clients must have the option to decide whether there's a legitimate need for an inspection fee, electronic transfer fee, processing fee or any other expense that's changed to them.

Last but not least, take your bankers to lunch once in a while. Once a quarter will do. The important thing is that you let them know that you want to do business with them and you need their help in putting your business together. Give them some of your business cards to keep on hand and some of your company brochures. Make it easy for them to recommend you. And most important, keep in contact with them.

Choosing a Good Broker

When you use mortgage brokers, you need to check them out thoroughly. Talk to a minimum of eight to ten people that they've secured loans for, and find out if there were any last-minute problems with their loans. If you hear of more than one instance where there was a delay in processing or securing the loan, then move on to another broker.

Mortgage brokers, unlike banks and savings and loans, don't actually have money to loan. For a percentage of the loan, they'll process the loan for your clients and find a lender to take it, or else they package it with other loans and sell them to a lending institution. Because of this, loans made through mortgage brokers have a higher rate of closing problems than loans made directly through banks or other institutions. There are two problems that occur more often than others:

1. The rate goes up at the last minute or there's an additional fee for processing the loan. If your customers don't want to pay the additional percentage or fee, the loan won't go through.

2. The loan conditions change, or the broker forgets to tell the client about some additional paperwork that the lender needs, such as copies of their last three years of tax returns or perhaps a copy their bank statements for the last 24 months. This usually comes up on the day the loan is set to close, thus delaying processing until all the requirements are met.

Even if a broker passes all tests to your satisfaction, you need to set down some ground rules for dealing with your company. The ground rules should be:

- No last minute rate increases
- No last minute charges of any kind
- No last minute paperwork surprises

If a broker can't guarantee you and your clients these considerations, then find another broker and save yourself a lot of grief. Brokers do have a say in these matters. They can set down guidelines for dealing with lenders the same as you can with them. They can refuse to deal with lenders who are uncooperative. A lender's money is only working for them when it's loaned out, not when it's sitting in their own account! So if a lender's main concern is thinking up new ways to get another nickel tacked on the end of a good deal, then let them extract that nickel from someone else.

Working for Bankers

If you do a good job of presenting yourself to your banker, you might find other opportunities opening up for you. Most banks have an R.E.O. department. R.E.O. is Real Estate Owned. These are properties that the bank owns, usually repossessed homes or buildings, and holds on its books. Bankers hate R.E.O.'s. They generally want to get them cleared off their books and turned into cash as quickly as possible. They need these properties cleaned, fixed up and sold. If you're interested in doing this type of work, let your banker know that you can move onto a "repo job," get it cleaned, painted and fixed as needed to go back on the market in a very short period of time. You could even go so far as to guarantee the bank that you'll respond to any request on R.E.O. property within 24 hours, and you'll give them a complete estimate for the repair and/or cleanup work to be done within three working days of your initial inspection. Of course, you'd better be sure that you can do this. With a little thought, you should be able to put a plan together for a quick response by your company. You could pick up some good business this way.

There's one thing that I'd advise you to do if you're going to take on R.E.O. work. Let the banker know that you're not going to "bid" the jobs cheap. If they want it done, and quickly, then they must be willing to pay a fair price for that service. If they insist on a "low bid" for the work, then pass on it. Some people (bean counters) simply don't understand the value of good, timely service. They'll jump at a bid that's $350 under your bid even if it means dealing with some flake that'll drag the job out three months longer than it would have taken you to do it. They don't see that saving $350 in repair work actually lost them $1,450 in interest that they could have earned had the property been on the market and sold sooner. All they see is the good deal they got on the "low bid."

Insurance Companies, Their Pay Schedule and Your Markup

Dealing with insurance companies is a form of masochism. If you've done insurance work, you probably already know all the stuff you have to go through to make a buck in that arena. There are whole books written on the subject of casualty repair — and several franchised systems that deal solely in that type of work. Let me briefly touch on the subject of markup as it applies to casualty loss and repair when there's an insurance company involved, either directly or indirectly.

Many, if not most, insurance companies will tell you that they'll only pay 10 percent overhead and 10 percent profit on any work you do for them. You must make a minimum 8 percent net profit on all jobs that you do. This sounds like it will work out — if you haven't read this book!

Let's look at what's wrong here. Using the numbers from our remodeling company back in Chapter 3, we can figure markup on a given casualty job. Here's what we have to work with:

- Overhead is 25 percent
- Job costs are 67 percent
- Profit is 8 percent
- Markup is 1.49

We've calculated the fire repairs and water damage for a home and estimated our job costs at exactly $10,000. Now in the eyes of most insurance adjusters, this will pencil out to be a $12,100 job, plus or minus a few dollars. They figure this by taking your job costs and multiplying them by 1.21, which is the same as multiplying by 10 percent, and then 10 percent again (overhead and profit).

What should our total sales price for the job be? Our actual job costs are $10,000. If you multiply that by our normal markup of 1.49, you get a sales price of $14,900. Big difference! What can we do to balance this difference out?

I've come up with a formula for a *job cost adjuster* that you can use for insurance work so you won't lose money on every job. To calculate the sales price of our job in numbers that we can live with and the insurance adjuster wants to see, we can take the following approach.

Divide $14,900 (our price) by $12,100 (the insurance company's price) and you get 1.2314. This number is our job cost adjuster, or JCA, as I like to call it. Now when we do our estimate, we simply multiply all of our job costs by our JCA (1.2314) to get our new adjusted job costs. We plug in these numbers on our scope of repairs form, or whatever name the insurance company has for it, and then add the approved 10 percent overhead and 10 percent profit that our friendly adjuster dictates.

Job Costs × JCA = Adjusted Job Costs

$10,000 × 1.2314 = $12,314

$12,314 × 1.10 × 1.10 = $14,899.94 or $14,900

Eureka! We end up with our correct sales price with the numbers that the adjuster wants to see. Everybody's happy!

There are a number of ways to work this out, but this approach is probably the easiest to use. You're playing with numbers, just like the insurance company when they insist that all contractors operate on a 10 percent overhead and 10 percent profit margin. We're all playing the same game, and it's a stupid game — perpetrated by the insurance companies. They know full well that the 10/10 percent system won't work for all contractors, yet they insist on this nonsense. You do have the option, however, of simply not getting involved in their game.

Some Advice on Casualty Work

Let me finish this section on working with insurance companies by passing along some advice that may save you some money and a lot of frustration.

First, be sure you know what the customer's deductible and depreciated amounts are before you ever do your estimate. If you can't get those figures from the owner or the adjuster, then be prepared to eat all or part of these amounts. If you know the deductible amount, you get that from the owner before you ever start the job. If you know the depreciated amounts, then you also get that entire amount from the owner before the job start, *no exceptions*. Of course, the homeowners won't want to pay those amounts. That's why you have to get the money before you start! Be very careful if the owner asks you to "build in" the deductible or depreciated amounts on your estimate. For an older building the depreciated amounts can run into several thousand dollars, and you might have trouble getting that by an adjuster. It's better to tell the owner that they must be responsible for any and all deductibles and depreciated amounts, and they must be paid 100 percent up front.

You should also find out the name on the insurance policy before you start any work. If it's just the name of the home or building owner, that's good. If it's an owner *and* a lender, you may have problems. The job may be in California, and the lender in Maine. When the job is done and the check is issued, the owner signs the check, you sign the check, and then the check is sent back to the lender in Maine. They, at their own leisure, will call for an inspection (which they'll ask you to pay for) to ensure the job was done right and the property is back in its original condition. When the inspector completes the inspection, he'll usually send the report to the insurance company instead of the lender. Five weeks later, after all the confusion is cleared up, you'll finally get your check. I've seen this process take up to four months. If there's a third party on the insurance policy, make the owner go get a short-term loan to pay the entire amount of the repairs. When the job is done, you get your money and the owner gets to wait for the check to come from the lender or the insurance company.

I also believe that contractors should charge for all insurance estimates. When you get a call for insurance work, even from former customers, tell them that there's a fee for insurance or casualty estimates. That'll immediately eliminate the people who are just looking for two or three estimates to give to their adjuster. Chances are these folks will pocket the money from the insurance company and try to repair the damage to their home or building themselves. I used to take a lot of these leads and found that they just weren't turning into sales. So I started requiring payment for these estimates. After a few months I found that the number of estimates I got called on to do was less, but my ratio of estimates to jobs sold on insurance work went from 1 in 32 to 1 in 3. It wasn't hard to figure out what the best approach was!

The last thing you should watch out for when working with insurance adjusters is the adjuster who'll wait until the last minute, when you've finalized your quote, to tell you that "We don't pay for job supervision." They always wait until you've completed your estimate because they know that having spent several days assembling your costs, you won't want to give up on the job. In most cases, what they'll ask you to deduct amounts to 6 percent of the total — what they feel is the typical cost of job supervision. If you do that, you'll eliminate almost all of your profit, even if you did pad your expenses. Most contractors panic at this point, take off the 6 percent, sign the agreement and lose most of their profit on the job. They know that if they refuse to make this adjustment, the insurance company will go out and look for other bids. Why is it so hard for us to walk away at this point? I don't know, but it really is.

Only you can decide how you want to handle a scenario like this — and believe me, if you work with insurance adjusters, it will come up. My advice is to be prepared to walk away. It's better to walk than to take a loss or work for next to nothing.

There are companies that do well working with insurance companies, but most don't. They realize too late that they can't run their business and pay their bills on the 10/10 percentage that the insurance companies pay. To be successful in the casualty repair business, you must first learn how to play the game. You must also have infinite patience, a high tolerance for frustration and be prepared to negotiate every estimate you do with an adjuster who'll always want you to cut your price.

Chapter 9

Justifying Your Markup

Some of the ideas we'll discuss in this chapter are also mentioned in other parts of this book. I'm including them here to give you a little different twist on how you look at these items, and to reinforce their importance to your company, its growth, and most importantly, its profitability.

If you're going to charge a fair markup on your work and a sales price that nets you a good profit, you have to be able to justify your price by providing the very best service you can. People don't mind paying more if they feel that they're getting good value for their money. How do you ensure this? You must build consumer confidence, and the best way to do that is through good work and good communication.

Communication

When you talk with people, give them 100 percent of your attention. If they're potential customers and you want their business, you must listen to what they tell you. This builds up their confidence in you and your company. Remember, they buy *you* first, *your company* second, and every other consideration, including the work to be done, comes in third or lower. In the mind of most customers, good communication equals value.

Phone Calls

Good service begins on the phone. That's generally your first contact with the customer, and it's usually the primary means of customer contact once you start the job. If you don't respond quickly and politely when a customer calls you or your office, you'll have problems with that customer. Be sure you have a cellular phone and a pager so that you can be reached anyplace you happen to be. Your superintendents should also have either a cell phone or a pager. In this age of electronic communications, there's no excuse for being out of touch.

Make sure you get all of your customers' messages and return all of their calls. This process starts with your office staff. They must take good messages, and pass them on to you or your job superintendent. Your employees will rise or sink to the level that's acceptable to you. Make sure they know what you expect.

There are several very good audio and videocassette programs available that demonstrate proper telephone etiquette. Even if you don't think you or your employees need them, buy one anyway and make it mandatory for everyone in the company to go through the program — including you and your field staff. When your staff has completed their phone training, test them by calling the office or paging them in the field every once in a while to see if they're responding correctly and within a reasonable time. If someone isn't answering the phone the way they should, give him or her a chance to go through the program a second time. If they don't catch on to the requirements, give them a transfer to the competition. You don't need people answering your phone who'll alienate your customers, or worse, drive them away before you even get a chance to talk with them.

Letters and Thank-You Cards

Letters of thanks and thank-you cards are great tools for promoting your business. How many other contractors send thank-you cards to their customers, suppliers and specialty contractors? I'm sure there are others besides me who follow through with a thank you, but for those who don't, here's something that should go to the top of your daily list of things to do. This is one other way to let people know that their time, effort, or consideration is important to you. Doesn't a sincere thank you brighten your day? A smile and a heartfelt "Thank you! You did a great job for us!" makes you want to do even more for people. Why not pass that feeling on to others? Let them know when you're pleased by something they've done for you or when you appreciate their consideration. You probably won't be able to measure the results of this effort on a chart or graph, but I guarantee it'll make a difference in your business. And, a thank you goes a long way in justifying your markup.

Pardon Me, Your Education Is Showing

While we're on the subject of writing, I'd like to discuss a subject that's near and dear to my heart. I may not be world's greatest speller, but I do make the effort to use correct grammar and spelling in all my written communication. The auto-correct feature in my word processor has about as many of my regularly-misspelled words in it as my spell-check dictionary has altogether!

As a rule, the spelling and grammar used by those in the construction industry is pretty bad. I get letters every day addressed to my software development company, Northwest Construction Software, and almost every one has misspelled words, incorrect grammar and missing punctuation. What kind of message does that send?

Now suppose for a minute that you've just been out to see a nice family about building a second-story addition onto their home and you want to thank them for considering your company for the job. So you write a thank-you note that says:

Deer Mr & Mrs Jones

Thank you for letting me come out and talk to the both of you. Were very intrested in the job and wood like to do it.

Sincerely

John Smith, Apex Construction

We all know that being able to spell or write well doesn't make you a good builder. But you're probably no better or worse at building than your competition. What counts, then, is the impression you give your customers. And how do you suppose these people are going to react to that kind of letter? Would you want to make a major monetary investment with a company whose spokesman exhibits this level of education? (And I do hope you recognized all the errors in that little note!) Compare it to the letter below:

Dear Mr. and Mrs. Jones,

Thank you for inviting me into your home to discuss your second-story room addition. We're very interested in doing this project for you. I will call you later in the week to confirm our next appointment. Again, thank you for your consideration.

Sincerely,

John Smith
Apex Construction

When you write to a customer, be sure you use correct grammar and always check your spelling. In my company, everyone checks everyone else's letters or other customer communications before they're sent out. Good business communication requires that you put your ego in your pocket and make sure that what you say to a customer in writing is said correctly. Your education is on display, and once a mistake is sent, you can't get it back. The impression is made.

Your grammar is even more important when you're speaking to customers. When you write, you have an opportunity to think about what you've written and check to see if it's correct. When you speak, once the words are out, you can't take them back! Think before you speak, speak clearly and don't use words that you're not sure of or can't pronounce correctly.

Your Employees

Take the time to talk to the people you work with on a daily basis. Give them a smile, and ask how they're doing. Inquire about their spouse, children, home, or hobbies. It only takes a minute, but it can make their day. Employees are dependent on us for their livelihood; that puts them in an insecure position. They need to know that you value them, both as employees and as people. If they do a good job, tell them. If they don't, remember, not everyone is perfect and even the best of us can have an off day. Maybe they don't do the job as well as *you think you could*. It's still important to get them to try and do their best. Build up their confidence and encourage them to keep at it. This takes a little time, a little interest and most of all, good communication. The results are employees who are loyal, hardworking and who'll stay with you for years — as long as they get their paychecks on time!

An employee who feels valued will pass that feeling on to others in the form of good will. They'll create a congenial atmosphere in their work area and will be pleasant and outgoing with your customers. A little interest in your employees can net you big dividends in the form of good customer relations.

Subcontractors, Specialty Contractors and Suppliers

Treat your subs and suppliers the same as you do your employees. In a sense, they *are* your employees. Talk with them. When they do a good job, say so. It's important to let people know their work is appreciated. It costs you nothing more than a phone call or a note, and it'll keep them coming back and doing their very best work for you on future jobs. If they do a bad job, or screw up an order (and they will), go out and discuss it with them. Let them know that anyone can mess up once in a while. Give them a chance to make it right — and 99 out of 100 times they'll not only make it right, they'll make it better! If you sub out most of your work, as I do, then you know that having a good relationship with the contractors and suppliers you work with provides huge benefits for your company in the form of good prompt service and dependable work.

Customer Response

There's been a lot written lately about getting customer feedback on how you did their job. This might be a good idea if you take the information they give you and use it to improve your work and your customer relations. However, if it's just going to be another form that you shuffle through the office without putting the information to use, then don't waste the time.

Your customers' responses will tell you the things that are important to them — what they liked or didn't like. It simply gives you a perspective on how they view your work, the way you and your crew did the work, and how your people conducted themselves on their job.

"There's an old saying in this business that says: 'Price too high, value too low.'"

The real benefit is that it'll tell you what to focus on to please your customers — and if you follow through, it'll help you justify the markup you're using.

There's an old saying in this business that says: "Price too high, value too low." Back in the days when I was selling for other companies and missed a sale, occasionally I would go back to the office and complain about the folks telling me our price was too high. (This was before I figured out how to get that nonsense stopped.) When I did this, one of the older, more experienced salesman in the company would almost invariably drag out that old saw, and of course, make my day complete!

Looking back now, however, I can see that there's a certain amount of truth to that old saying. You must build value into your jobs to get the markup or the sales price that you need for your business to survive. In other words, there's a direct correlation between what you can charge and the value you produce. The price is only too high if the value is too low. People will pay well for good work.

Classes, Seminars and Conventions

As we have already said, and will say in other parts of this book, it's the wise and prudent contractor who continues his or her education. You must keep up with the latest in building materials and methods, and be sure you're aware of new business trends and better ways of conducting your business. One way to do this is by attending classes, seminars and conventions held for those in your field. Why is this important? Because keeping up with all the latest innovations in your field provides you with one more tool to justify your markup.

Time for another gut check. Did you have an Operating Capitol Reserve Account in place before you started reading this book? Had you ever thought about setting one up? Did you know how to factor the cost of that account into your markup? These are some of the things that you learn about when you get involved in the educational programs that are available to you — and that we all need. No one is going to come to you and tell you what's new — except maybe your customers. Remember, *they* like to keep up with the latest trends. Don't become one of the unconscious incompetents who stumble through life without changing. Just because your father did it this way, and your grandfather did it this way, doesn't mean that a better way doesn't now exist. If you don't get out and talk with others, attend classes, seminars and conventions, you'll never know what you're missing — and that can hurt you.

Chapter 10

Your Employees and Your Markup

There are a few basic things that we need to talk about regarding your office staff before we talk about their schooling in markup. This will apply to all the staff, regardless of the number or whether or not they're related to you.

Training

First, every new employee needs some training for the job they're hired to do, regardless of what that job involves. I think one of the biggest mistakes the owner of a construction company can make is to assume that a new employee knows everything there is to know about working in an office. You should have a checklist of things to cover with each new employee and go through that list with them item by item. Even if they appear to have many years of office experience, you have to ask yourself, "Does this person have 15 years of experience, or 1 year of experience repeated 15 times? Did anyone ever train them in the proper way to deal with customers, or did they just pick up good and bad habits through the years?" Don't take chances. Tell them about your company, what you do, how you do it, and what you expect from your employees.

At some time during their first 90 days with your company, everyone you hire to work in your office should go out with you or another salesperson on a sales call and then follow that job from start to finish. That means they go on all callbacks, to the contract signing, and then several visits to the job site to see the actual construction process through to its final completion. When they've done that, they'll know how your company works and they'll be a much better employee — and far more knowledgeable when they answer the phone. If you think this isn't a necessary part of their training, pick someone in your office that hasn't gone through this and ask them a few questions about what other people in your company do. You'll be surprised by their lack of knowledge. It's really unfair to put them in the position of answering customer questions until you've given them the background to do the job properly.

Chapter 10: Your Employees and Your Markup

Productivity

The second problem we need to address is the non-productive employee. Everyone on your staff should be productive. By productive I mean that every individual should have all their efforts aimed at bringing money in the door. Most construction companies are overly employed. They have more people working for them than they need to get the job done. Look back at the charts called Volume Built per Employee on pages 43 and 58 in Chapters 2 and 3. If your production per employee doesn't come within 10 percent of what these charts show, then it's time to lean up and clean up.

So how do you change your bookkeeper from an overhead entity to an employee who does bookkeeping and also contributes to the income and profitability of your company? One way is to make sure that your bookkeeper, as well as everyone else in the company, has company business cards to give out. I would think that the minimum number of cards that each person in your company should give out in any given year would be 500 cards. That's just one box, or two cards a day. That should be a fairly easy task to perform for almost anyone. I give out an average of 50 cards a day throughout the year, so two cards a day shouldn't be an unreasonable request of your employees.

Your staff can be your representative at various community organizations that they belong to, such as the Kiwanis Club, the Lions Club, the Chamber of Commerce, the Elks, the local PTA, youth athletic organizations, church groups, and so forth. They'll be happy to do that for you as long as you in turn contribute financially or constructively to the needs of their group or organization. There are always ways that your company can help, whether it's buying an ad in the school newspaper, sponsoring a youth soccer or baseball team or building booths for a community fund-raising project. Anything you do for others increases your exposure and your good name in the community.

"Non-productive employees do just one thing — raise your overhead, and that in turn raises your markup."

Take members of your staff along to construction association meetings when you attend. That'll give them a chance to meet others in the business and see who your competition is. Have your office staff spend one day a month working in your showroom or in your model homes. You can also have them participate in your booth at local home improvement and remodeling shows. Give them the opportunity to rub elbows with the buying public. It's easy for them to get wrapped up in their area of the business and forget what the company is really about. They need to see the big picture and be reminded that the company's focus should be on the needs of the customer and not just the pile of paperwork that their job can become.

As the owner of your business, you are charged with getting the very best from each of your employees. That means getting the maximum production out of them. I'm sure it won't surprise most business owners to hear me say that I think people are basically lazy. While they may have good intentions, the average employee will still only produce what's required to keep the boss happy and their paycheck coming in. Occasionally, you'll find that rare individual who "gets the job done no matter what it takes," but most people just get by.

Of course, employee attitudes are often just a reflection of the attitude of the company's owner. I've been absolutely amazed at the attitudes of some contractors who've called me out to help them get their business back up on its feet. You wouldn't believe the number of times I've heard "Well that's the way we've always done it here, and it's worked OK so far!" What can I say? I'm sure you've guessed what I say: "If it's working, what am I doing here?" Keep your attitude fresh. Be sure that you don't get so set in your ways that you're unwilling to look at new ideas or new approaches to your work. It doesn't make any difference if you think your way is better, just be willing to look at the new approach objectively and give it a try — even for a little while. If you're willing to do this, your employees will follow your lead.

Non-productive employees do just one thing — raise your overhead, and that in turn raises your markup. Unless you're completely oblivious to the effects of an increased markup on your business, you know that you must keep everyone in your company focused on bringing in new customers and then providing those customers with the very best service available.

Company Meetings

While we're on the subject of productivity, let's talk about company meetings — one of the least productive employee/management activities that I can imagine. For those of us in the construction business, company meetings are not only a waste of time, they're a form of employee abuse! How much monotony can you expect an employee to put up with?

In all the years that I've spent in this business, I can count on one hand the number of meetings that I've sat in on that were worth the time invested. Most were a colossal waste of time. I can hear the screams of outrage and disbelief from those owners and managers who think that their meetings are a productive part of their businesses. But, if you want the real truth about your meetings, make up a questionnaire for your employees (one that they can respond to anonymously) and ask them what they really think. They'll probably tell you the truth about your meetings — just be prepared

Chapter 10: Your Employees and Your Markup

for a real eye opener. It shouldn't take more than four or five questions to get the gist of their feelings. Here are some suggestions:

1. Do our meetings start and finish on time?
2. Are these meetings the best use of our time?
3. Do we accomplish the agenda of our meetings?
4. Do you feel you can offer comments and suggestions at these meetings that will be taken seriously and that someone will follow up on?

The following guidelines are used successfully by several companies for their meetings and they make a lot of sense to me.

- No chairs allowed in the meeting room. The meetings should be conducted with everyone standing. That'll guarantee that the meetings will be short. (Studies have shown that standing will cut the meeting time by at least half.)

- The meeting agenda must be outlined, double-spaced, on a single sheet of paper. Each person attending the meeting should receive a printed copy. This provides a clear set of goals to accomplish, and keeps everyone focused.

- The individual who called the meeting should prepare the outline, not someone on their staff. Without exception, whoever called the meeting must be in attendance.

- There should be a definite time limit set for the meeting when it's scheduled. The meeting must then start and finish on time, no exceptions.

If you hold weekly sales meetings, try changing to every other week, or once a month. Before you make the change, check your sales-to-leads ratio for several previous months. If that ratio begins to drop, say from one sale for every four appointments to one for every five or six when you change your meeting schedule, then your meetings were probably having a positive effect, and you should keep them at once a week. If the ratio doesn't change when you change your meetings to every other week or once a month, then what impact are your meetings having? The answer is none. If your productivity increases with the change, and your sales go up from one in four appointments to one in every two or three, then perhaps your staff's time is better spent on sales calls than in meetings!

If you've clearly defined sales goals for your company, and each salesperson has their personal goals set, then you've given them all the direction you can give them. At that point, either they're motivated to go out and fulfill their goals or they aren't. If you really want to help them, go with them on a sales call and keep quiet and watch how they do their job. If they're having a problem, you'll be able to pick up on it. After the sales call (never during it), review their approach and offer them some tips or suggestions that they can use to improve their sales techniques. Now that's a productive sales meeting — it's specific, to the point, and truly helpful!

Schooling on Markup

Many business owners are reluctant to talk to employees about the financial aspects of the company in general, and specifically how the company charges for their work. That may be a good idea. With the rate of employee turnover in the construction business, today's helpful employee may be working for your competition tomorrow! So how much should you tell your employees about how you charge for your work?

If you have employees, then you should have a training program in effect for them. Part of that training should include company policies and a brief discussion on the economics of your business. You can certainly instruct new employees on the financial aspects of your business without giving away the store, so to speak. However, depending on their role in the company, some employees will naturally need to know more details about your company finances than others will.

Training programs for employees aren't easy to write. What to include, what's relevant to their position, and what to skip over can be difficult to determine. As a general rule of thumb, I'd say that if they're involved in writing contracts or additional work orders, or processing either one, then they need to know all about your markup and how you arrived at the markup you use.

Your field crew, carpenters, and outside helpers should know your approximate volume of business sold, built and collected each year. Their focus will generally be on their benefits and how they're paid. They'll seldom need to know your company markup unless they write up additional work orders — and usually that's something that only your superintendent will get involved in. You'll need to show your superintendent how to calculate markup in the same manner that you train your sales staff.

Your office accounting staff and your sales staff need to know all about markup and how it's figured for your company. They need to understand how important markup is to the overall success of the business. The better their understanding of this key element, the more help they'll be to your business. This is especially true of your sales staff. The more thorough their training is, the less tempted they'll be to try and figure out ways to cut the markup on the jobs they're selling. And trust me, salespeople are born looking for ways to cut prices! It's part of their nature. So if they thoroughly understand what it costs to run your company and why your markup is necessary, then they'll be far less likely to spend time trying to cut the sales price of the work that you're doing.

You have to be very clear about the markup that your sales staff must use. This should never be a topic of debate. You're not running a democracy; you're running a business. Once you've determined what your correct markup is, you must write clear guidelines about how it's to be applied. Keep a tight reign on the markup used by everyone in the company who does any selling, including you. You can't expect your sales staff to use the correct markup if you don't. Never think for a minute that they won't find out if you cut your markup — and when they do, you'll instantly lose credibility with your staff.

However, there may be an occasion when someone feels there's a valid reason to make adjustments to your markup on a particular job. This should be the option of the owner or general manager, and only the owner or general manager — never a member of the sales staff. They must bring the suggestion to you for your consideration. My recommendation is to make it so tough to get an OK to change the markup that most salespeople will decide it's easier to just go ahead and use the correct markup than spend time trying to get it lowered.

You can test your employees' understanding of markup by having them do some sample problems. There are several that you can use in the section called *Problems to Solve* in the back of this book. If you've taught them well, these markup problems should be easy for them to solve. If you don't feel up to the task of training your staff yourself, you can make copies of the key chapters of this book for them to read, and then give them the problems to solve at the end of the book (without the answers) and see how they do. Make sure they stay with it until they can do all the problems correctly. Never send anyone out to sell anything for your company until they thoroughly understand how to properly calculate your company markup. After all, good training and a good understanding of markup are key elements in any company's success.

Chapter 11

Listening to the Experts

As you move along in your field and begin to appreciate the advantages of expanding your job knowledge and your education, you'll invariably run into individuals who consider themselves "industry experts." (I've mentioned them once or twice before.) They're often lecturing or teaching the seminars that you attend. I can hear you asking right now, "Doesn't Michael Stone lecture and teach seminars?" And to answer you, "Yes, I do. But I don't fancy myself an industry expert."

If someone were to call me an industry expert, I'd probably be more insulted than flattered. While I do have some strong opinions, and I've spent a lot of time and a ton of money learning all this stuff I go on about, I doubt that I'll ever know enough to really be an industry *expert*. An expert is someone who has "special skill, knowledge or experience representing mastery of a particular subject" (Webster's Dictionary). So, even though I have a good understanding of the subject of markup and profit, that hardly qualifies me as an expert on the construction industry. This is an enormous business, making up almost 10 percent of the U.S. gross national product. How can one person wear the label "industry expert" in an industry involving so many different trades, and each with so many facets? They can't — except in their own minds and the minds of a few magazine editors.

Recognizing an Expert

How do you decide who is an expert and who's just a lot of talk? One way is to listen carefully to what they say. When you find an individual who's noisily proclaiming his expertise, and there are plenty of them out there doing that, ask him or her some specific questions about the phase of your business that you know best. You'll quickly discover if the label "expert" fits. Find out if this person is still working in construction. It's one thing to stand up in front of a group of folks and tell them how to do something. It's quite another to have done it in the last 30 days. If they aren't in the business on a daily basis, they're out of touch. It's just that simple.

Several years ago, I attended a seminar at a national builder's convention. The lecturer was the editor of a national magazine and his topic was on how to sell remodeling. It didn't take me long to realize that he was totally out of touch with the remodeling industry. I wondered, in fact, if he had ever gone out and sold a remodeling job. It was obvious to everyone in the room with half an ounce of sense that this "industry expert" wasn't! The worst part of it was they had paid good money to sit and listen to this guy.

After that experience, I decided to call and check into the lecturer's training and background before I paid to attend another seminar. I've found this to be a very good idea. One seminar I called had put out an advertising flyer that said the lecturer was "uniquely qualified to teach this class." When I checked into this person's training in estimating residential remodeling, I found that he had never estimated or even built a remodeling job — but he had a lot of experience in real estate and new home construction. What was "unique" was how that qualified him to teach a class on estimating remodeling jobs! It's amazing what some companies will pass off as expertise just to make a buck out of those of us in the industry.

"How can you spot the genuinely knowledgeable people in this business? They're the ones who want to talk to you."

Ask a lot of questions before you sign up for any seminars or classes, and once you're in the class or lecture, listen closely to what's being said. Before you take anything at face value, find out just how much this "expert" really knows. Watch their posture, watch out for statements in which they use a lot of words but offer no explanation. See if they answer questions directly or skirt around them. Their answers will usually tell you if they really know what they're talking about or if they're simply repeating ideas that they've heard from other people. Sometimes these experts pick up on an idea at a meeting or convention somewhere and think it sounds good. But before they've tried it for themselves, they climb onto the bandwagon and go around the country telling everyone about it. It sounds good, they take all the credit, but do they really understand the concept and whether it would or wouldn't work for you in your field? Ask tough questions and listen carefully to the answers. Many times you'll find that you know as much or more than they do.

How can you spot the genuinely knowledgeable people in this business? They're the ones who want to talk to you. They usually make themselves available so they can answer your questions directly. They're quiet rather than boastful, easy to talk to, they practice good eye contact, and most importantly, they're still active in the business that they're speaking on. You can tell when you talk to them that they've been there, done that, and done it recently. They'll ask you questions too, and then they'll listen to what you say. It's a two-way conversation for them. They may have some strong opinions, but they also have open minds. When you find someone like that, listen and learn.

Increasing Your Expertise in Your Field

How do you become knowledgeable enough in this business for others to seek your counsel? Assuming that you have a desire to be of help to others in this business, there's a formula promoted by Zig Zigler, Brian Tracy and others in the self-improvement industry that's generally accepted as the best one to use to reach that goal. It's this: Read about your business at least one hour a day, every day. If you do that, you'll become recognized for your knowledge locally within one year, nationally within three years, and internationally within five years. That represents 1,825 hours of reading over that five-year period. If you want to attain your goal more quickly, you can increase your reading time. If you double your reading time to two hours a day, you can cut the time it takes to reach your goal in half. However, don't spend your time learning just to impress others with how much you know. Be a student of the business, and learn how to apply what you studied to make your business grow. You can impress others more by your example than by anything you say.

Suppose, after a while, people begin coming to you and asking you to do seminars, classes, or other teaching stints. As I learned the hard way, doing classes, seminars, and speeches at conventions is great for the ego, but it seldom puts money in the bank, or food on your table. As a matter of fact, it'll often end up costing you money. If your ego gets out in front of your common sense, it can get expensive. Not only will you have to spend money getting to and from the speaking engagement, and many times paying for your room, meals and transportation while you're there, but you'll be taking valuable time away from your business. That's time you should be earning money, not spending it. In short, you're not staying focused on what you set out to do.

If you don't want to become an overnight wonder, but you do want to expand your general knowledge, here are some basics that you can easily accomplish during the next year:

- Read at least six books on business management, and at least the same amount for personal growth and enjoyment.
- Attend at least three one-day seminars on some form of business management, sales or other business-related topic.
- If you didn't graduate from high school, start taking some night-school courses to get your high school diploma or a G.E.D.

- If you did graduate from high school, think about taking some college courses or getting a four-year degree. You can enroll in a program at your local community college and go on from there. Many colleges and universities also offer certificate programs in various areas of construction.

Organizations and Business Associations

What about getting involved in the hierarchy of the trade associations? While I'm all for belonging to and working with an association *in your field of business or area of interest*, the big question I ask myself before making a commitment to any special involvement is, "Will it put money in my bank account?" Here again, it usually won't. If the involvement takes you away from your business, it's going to cost you money. This also applies to "donating" your time to serve as a committee chair in a local or national association. It may stroke your ego and bring you recognition, but too much time away from your business can ruin you.

I can think of a couple of associations that I've belonged to over the years in which you were almost guaranteed to have your own business go bankrupt within the year if you took on the position of president or chairman. The demands of the position were so time-consuming that your own business had to be relegated to second place behind the needs of the association. Bad plan! Seems to me like something is wrong here! The organizations are supposed to help their membership survive the rigors of the business, not drive them *out* of business. If you find yourself trapped in one of those positions, walk away before it's too late. Resign, and get back to doing what you should be doing — focusing on your own business.

Networking and Networking Groups

I've never belonged to a networking group, other than one for generating new leads for my business. There are now several of these networking groups for those in construction, and I think they're good for the industry. The advantage, as I see it, is that you can get a variety of opinions about how to do something from individuals who are doing it themselves. You can share first-hand experience with others in your field.

These groups meet at a neutral site or place of business and discuss topics that are of concern to all their businesses, or they may assemble at one members place of business to discuss the particular problems of that company. Generally, individuals who put effort into building up their own businesses and then branch out to help others in their group are pretty good at getting to the heart of a business and its problems. They can probably give some good advice and direction to any business they hold under their magnifying glass for a day or two.

If you're interested in joining a network, you can probably get information on local groups from your association. If they don't know of any local groups, a national association office might be able to direct you to one close by. You might also get information from talking to others in your business field. Often the best information comes from those who are involved in the groups themselves.

Be sure you check into the background of the organizer of any network group that you consider joining. Does this person know as much about your field as you do? Is he or she working in a construction-related field every day? Do they or did they have their own construction company? If they no longer have their own company, what happened to it? Check out all these questions and make sure that you get the answers. I would also ask for a list of some of their old customers that I could call and talk with. And, I would contact the Attorney General's Office in the state where they operated their business and find out if there were any complaints filed against them or if they have ever filed bankruptcy. If you're going to pay someone to join their network (and believe me you will have to pay), then you certainly need to know something about the background of the person taking your money. Don't you expect your customers to do the same with you?

Joining a networking group is no different from joining any other organization or group. Check it out first. Don't waste your time getting involved with any group unless it measures up to your expectations and you can get as much out of it as you're expected to give.

Hiring Consultants

There's a definite time to hire a consultant, and the very best time is when you know you have a problem but you're unsure what to do about it. What you do know is that you can't pay all your bills, and you need help. If you own the company, then you're the one responsible for making the decisions that got your company into the financial position you're in today. Now you have to make another decision. If you make the right decision, you survive. If you make the wrong decision, you become another sad statistic.

I was once told "Anyone who can afford a consultant, doesn't think they need one!" That takes you right to the heart of the matter of why so many contractors fail. If you're absolutely sure that you don't need any outside help, then I invite you to ask yourself the following question. Am I making the minimum 8 percent net profit that I should be making? Careful here, this is a test to see if your ego is in your pocket where it should be while you're running your business. If you can honestly answer yes to that question, I'm delighted for you. If you aren't making that 8 percent, then you need to do the following:

1. Review this book.

2. Call in a consultant, even if you're sure you don't need one.

A consultant is one "expert" that you can't afford not to listen to. To find a good consultant, check with the association that you belong to. The director of the association should be able to suggest a few people who can help you. If you don't belong to an association, then call some of your friends in the business and ask them to suggest one or two people who could advise you. A consultant doesn't have to be an expert, but he or she does need to be successfully running a business similar to yours.

"I was once told 'Anyone who can afford a consultant, doesn't think they need one!' That takes you right to the heart of the matter of why so many contractors fail."

When you get the names of several people, then you can begin checking them out to come up with one who knows what they're doing and who'll also be willing to help you. You may have to make a few calls and expend a little effort to find someone you can work with. Take the same approach in locating a consultant that careful customers take in locating a good contractor to build or improve their home. If you do that, then you'll be hiring someone that you have confidence in and whose advice you feel you can trust.

The bottom line is this: Find someone you can really rely on for help, and when you need that help, get it.

Some Good Advice

Now I'd like to give you a brief summary of some of my best advice. This comes not from an expert, but as the accumulated wisdom of some very good business people that I've met over the years. So pay attention friends, here it is.

My Recommendation

Here's some potentially-controversial advice, but I've found it good nonetheless. I've observed over the years that many successful construction companies have women making most of the major business decisions. I would suggest that if you're a married man and you own a construction business, get your wife involved in your business — at the very least as your consultant. I was brought up to believe that women were the emotional ones, not strong enough to make tough business decisions. That simply isn't true.

It may be that when it comes to this business, women deal from logic and men from emotion. I'm certainly no expert here. But I've found that when it's time to decide, men tend to let their feelings *and their egos* get in the way. If I've decided I want to sell a particular job, I just hate to let it go, even if things aren't looking promising. I want to keep looking at it and see if I can turn it around. I hate to fail — which is how I see it. My wife Devon, on the other hand, has no such problem. "This job's going to lose us money — forget it!" Women seem to have the ability to look at a situation, size it up and make a decision based on the facts. They can see the big picture. Men often "let their ego get out in front of their common sense!"

If you doubt this, ask an attorney to compare lawsuits generated by businesses, fifty owned by men and fifty owned by women, and tell you which initiated more lawsuits over any given period of time. You'll discover that for every lawsuit generated by a female-owned business, there are at least three generated by male-owned businesses. Men know just as well as women that the only winners in most lawsuits are the lawyers, but men want to duke it out anyway, even though they know it'll cost them more in the end. Women have more sense. If you're lucky enough to be in a partnership with one, turn to her for advice, then follow it.

Know When To Say No!

You'll know that you've reached "business maturity" when you're able to walk away from a potential job. There are lots of reasons for not taking a job, but the most compelling is the one that tells you that you're not going to make any money on it, or worse, that you could lose money on it. Let's look at a few situations where this is most likely to happen.

The Low Bid

When the customer, whether it's an individual, a lending institution, a governmental agency or an insurance company, tells you they're after the low bid, just walk away! This is a no-win situation for most companies. There are only two ways that this ever works out in favor of the contracting company:

1. If you calculate your estimate, know your numbers are right, and you can use a markup of your job costs times at least two or more and still get the job, then you probably will make some money. If you can put together a bid like this, which is doubtful, go for it. Anything less doesn't justify the amount of time you'll spend putting a "low bid" together.

2. Large companies can often enter into "bidding contests," get the bid, and still seem to not only survive, but thrive. Those companies know to the penny what a job will cost them to build, and believe me, they're getting their full markup on those jobs. But they're operating at a level far beyond most of the contractors we know. They're dealing in volume sales. Their overhead as a percent of their total sales price for their jobs is very small, thus their markup will be smaller than yours. They know their overhead numbers and they have the discipline to develop a budget for a job, stick to it like glue, and make it happen. If you're ever in that position, you can outbid the little guys and still make money, too. But until that time, stay away from the "bidding" game. In general, low bidding is suicide for your company.

Here's something I recently heard that sums up how I feel about low bids: Competitive bidding is like shooting yourself in the foot! You drive down prices and make it harder for everyone in the business to make a profit.

Problem Customers

When customers have their own ideas on how a job should be done, and know just enough to be dangerous, walk away. I've found over the years that if a customer has strong opinions on how you should do your work and they're difficult to deal with before the contract is written, they'll be even tougher after the job is started. If they have their own agenda, say no.

Another warning sign should pop up in your brain when a customer asks you to just send your bid or call with a bid rather than meeting with you face to face to discuss the job. When this happens, you know that they're probably shopping for a low bid and they're afraid that if they see you they'll have to make a decision to buy from you. Don't waste your time with this type of customer. If you ask enough of the right questions when you first speak with people, they'll tell you everything you need to determine if they're a serious customer or not.

Negotiating Price

When a customer insists on negotiating the price of a job, walk away. You should only negotiate what you're going to do, how you're going to do it, when and where you're going to do it, but never, never, never how much you will charge for your work. If you give in on this issue up front, they'll be trying to negotiate your price down throughout the job. Just say no, and be prepared to walk away.

Time Management

You and I wake up each morning with the same amount of time to get things done. What you do with your time is the key factor in determining your success in the construction business, or in your personal life, for that matter. You can make all the rationalizations about your life that you want, but the bottom line is that the way you manage your time determines how much wealth, or misery, you'll accumulate. When you take responsibility for that fact, then you can begin to make your life better. That's the first step in better time management.

There are dozens of books and audios and videos available to help you solve time management problems. Any effort that you make to improve your time management is time well spent. How do you know if you need to improve your time-management skills? Have you used those lame and tired excuses: "I'm so busy, it just got away from me" or "We're just so swamped" as an excuse for missing an appointment, or not getting back to a customer in a timely manner? Remember what I said earlier? These are just other ways of saying, "I'm too disorganized to do my job well." You're no busier than anyone else is!

Get yourself a daily planner or organizing book of some kind that you keep with you at all times. *Day Timer* is a fairly common example of this type book. It should have a place for names, addresses and phone numbers as well as your daily, weekly and monthly schedule. If something is important to you, it should be in there. You can also get an electronic notebook that'll do the same thing that you can do with a manual planner, except you can program in reminders and updates.

Focus

Last but not least, here's the single best piece of advice in this whole book. You must develop the ability to focus on what you're doing if you want to make money. By focus, I mean putting all of your thoughts and energy towards accomplishing a given task, no matter how large or small. Focus on the things that you can control, and ignore the things you can't control. Don't get hung up on worrying about things that happened in the past or that could happen in the future. Take control of the things that you can and follow through on doing what you can to make things better. It's hard to let go of the "what ifs" or the "should haves." I've been working on this for years. Here are the rules I follow:

1. Focus on what you're doing
2. Ignore the things you can't control
3. Stick to your daily schedule

I've given you some good ideas to work with. What you choose to do with them will have an effect on the balance of your business career and your life. Remember, what you do with your time is what determines your lot in life. Make your time count starting right now, and *STAY FOCUSED*. As my hero, John Wayne, used to say, "That's some good advice, pilgrim!"

In Conclusion, May I Say Thank You . . .

The next section of this book has the problems, with their answers, that I promised you. Please go on and work them through now, while the information is fresh in your mind. In the appendix, following *Problems to Solve*, you'll find blank forms that you can use in your business, and a list of recommended reference materials that I have found particularly helpful.

You've heard before that business, like goal-setting, isn't a destination — it's a never-ending journey. I love the construction business. It gets in your blood. It's a great business, and the best part is working with people like you. There are none better in any business or profession. I've tried to make the path a little less bumpy for your trip, perhaps easier to travel than it's been for others I've known. I'd like to believe that I have. I sincerely hope you will do well in your business endeavors, and I wish you, your family and your business the very best, always.

Many of my friends and acquaintances have contributed to this book, and for that I will be forever grateful. To them I say, "Thank you for sharing your experience and your knowledge!" You have not only helped produce this work, but you have contributed your input in helping make the road to success in the construction industry just a little smoother for us all.

Problems to Solve

I've included the following problems for your review. They'll provide a good measure of how well you've assimilated the material in this book. The problems are fairly typical of ones that most contractors face, regardless of their time in the business. Everything that you'll be working out has been covered at least once in this book, so if you have any trouble, go back and reread Chapter 3 on markup and Chapter 7 on margin and break-even point.

This is another gut check. This is where we'll see if you have the determination to follow through and do these exercises. Some of you, I know, will just skim over them and not be willing to spend the time to solve for the correct answers. As in all things in life, you'll only get out of this what you put into it. If you "peek" ahead at the answers before you work the problems, you'll never know if you fully understand the information that's provided here.

It's worth investing the time and effort to do the problems. Give this the same effort you'd give a problem that's come up in your company. Then when the problem actually happens — and it will — you'll already know how to solve it. If you get stuck, stop for a while, even a day or two, and then come back to the problems. The answers will sometimes come more easily after a little break. Don't worry if it takes you a while to figure out each problem. There's no pressure — you're testing yourself. The purpose is to make you think, and to help you see if you understand the logic behind applying markup in your own business. In most of the problems, I'll be your silent business partner, so the problems are all written just like the ones in the book, with us working together. The last two problems are single ownership problems, with you handling the operation on your own. This is an open-book test. You may refer back to the problems we've already done any time you think you need help.

NOTE: In all problems, Sales, Sales Volume, Volume, and Volume Built will be considered one and the same.

Problems to Solve

Problem 1: Getting Started

After working for other contractors in both new home construction and remodeling for several years, we've decided it's time to start our own business. We both have our own tools and we've put some money aside to live on until we get the business to a profitable state. We have our license, insurance, bond and all the rest of the required trappings that make the bureaucrats happy.

We have a certain amount of solid business facts that we've been able to draw from, but there are a number of items that we've just had to guess at when making our projections. We're not hiring employees, but plan to do everything we can ourselves and sub out the rest. We've projected that, during our first year, we can sell, build and collect on 28 residential remodeling jobs, with an average sales price of $9,600. After researching our potential budget, we've estimated that our overhead each month will come to $6,185. Our goal is a 10 percent net profit. Our advertising has been out for about two weeks and our phone has started to ring. We're ready for our first sales call.

Questions:

A. What should the markup be for our new company?

B. What's our margin and percent of margin for our new company?

Answers to Problem 1: Getting Started

A. What should the markup be for our new company?

1. Our first year projected sales volume is $268,800 (28 × $9,600).

2. Our first year projected overhead expense is $74,220 (12 × $6,185).

3. Our first year projected net profit is $26,880 (10% × $268,800).

4. Our job costs are $167,700 *(Total sales − Overhead − Profit = Job Costs* or $268,800 − 74,220 − $26,880 = $167,700).

5. Our markup is 1.61 *(Sales ÷ Job Costs = Markup*, or $268,800 ÷ $167,700 = 1.60286. Remember to always round up.)

Answer A: Markup is 1.61

B. What's our margin and percent of margin for our new company?

Margin:

1. Our *Margin = Sales Volume − Job Costs*, or $101,100 ($268,800 − $167,700).

   For your information, you can also use the formula, *Overhead + Profit = Margin*. The result is the same: $74,220 + $26,880 = $101,100.

Percent of Margin:

2. Our percent of margin is our *Margin ÷ Sales*, or 38 percent ($101,100 ÷ $268,800 = 0.376).

   Of course, these figures may change once we get some real experience behind us. The only way to really know for sure what things will cost and how much money we can make is to go do it, then monitor and analyze the results.

Answer B: Margin is $101,100.
Percent of margin is 38 percent.

Problems to Solve

Problem 2: A Fourth-Year Business

We own a construction company that specializes in building new custom homes. We don't do spec building; we just build pre-sold homes for customers. We're now in our fourth year of business and have done well to date. It's July 1, and in doing our quarterly review, we have gathered the following information about our company.

Our sales to date for this year are up 13.45 percent over last year at this time. Our projections for this year show a 10 percent increase in overall sales for the year. Last year we sold, built and collected on seven new homes for a total sales volume of $1,472,396. During our last 10 quarters in business, we've been within 2 percent of our projected quarterly goals for jobs sold, built and collected. Last year we made an 8.79 percent net profit. The year before we made an 8.21 percent net profit. Our overhead last year was 7.53 percent; this year it's running about 1.5 percent higher.

Questions:

A. What's our projected volume in jobs sold, built and collected for this year?

B. What's our projected overhead expense for this year?

C. What's our projected profit for this year?

D. What will our projected job costs be for this year?

E. What should our markup be for the remainder of this year?

F. What's our projected percent of margin for this year?

G. Approximately when will we reach our break-even point for this year?

Answers to Problem 2: A Fourth-Year Business

A. What's our projected volume of jobs sold, built and collected for this year?

1. To find our projected sales for this year, take last year's sales of $1,472,398 and multiply them by our projected 10 percent increase.

2. $1,472,396 × 1.10 = $1,619,635.60.

Answer A: Projected sales are $1,619,636.

When we work our numbers for future projections, such as projecting our sales for the remainder of the year or for future years, we must rely on our previous company history. In the past 10 quarters we've been within 2 percent of our projections. We'll assume, therefore, that our original projections of an increase of 10 percent for the year are accurate, and continue to use that volume for our projections through the end of the year. While we're 3.45 percent ahead of what we were doing last year at this point in time, and hope to continue at that pace, *our history*, not our hopes, unfortunately, must be the guiding factor here.

B. What's our projected overhead expense for this year?

1. To figure our projected overhead expense for this year, we multiply our projected sales volume by our estimated overhead percentage of 9.03 percent (7.53 + 1.5 = 9.03).

2. $1,619,636 × 9.03% = $146,253.

Answer B: Projected overhead for this year is $146,253.

While there's still time to make an adjustment in our overhead spending from now until the end of the year to get the 1.5 percent increase reduced, we have to assume for now that the increase was budgeted, and go with that number through the end of the year.

C. What's our projected profit for this year?

1. To figure our projected profit, we multiply our projected sales volume of $1,619,636 by our projected profit percentage of 7.29 percent (8.79% − 1.5% = 7.29%).

2. $1,619,636 × 7.29% = $118,071.

Answer C: Projected profit for this year is $118,071.

Problems to Solve

While we don't know exactly what percent of net profit we're making this year to date, we do know that our overhead is up by 1.5 percent. That doesn't necessarily mean our profit has gone down, but we'll assume the worst until we prove otherwise at the end of the year.

D. What will our projected job costs be for this year?

1. Our job costs are our *Total Sales − Overhead − Profit*.
2. $1,619,636 − $146,253 − $118,071 = $1,355,312.

Answer D: Projected job costs are $1,355,312.

E. What should our markup be for the remainder of this year?

1. Our markup is determined by dividing our total sales by our job costs.
2. $1,619,636 ÷ $1,355,312 = 1.195.

Answer E: Projected markup should be 1.20.

F. What's our projected percent of margin for this year?

1. First we need to figure our margin: *Sales Volume − Job Costs = Margin*.
2. $1,619,636 − $1,355,312 = $264,324.
3. *Margin ÷ Sales Volume = Margin Percentage*.
4. $264,324 ÷ $1,619,636 = 0.1632.

Answer F: Projected percent of margin is 16.32 percent.

G. Approximately when will we reach our break-even point for this year?

1. To figure our break-even point, we'll use the formula:

$$Break\ Even = Total\ Overhead \div \frac{(Total\ Sales - Job\ Costs)}{Total\ Sales}$$

2. Break even = $146,253 ÷ $\frac{\$1,619,636 - \$1,355,312}{\$1,619,636}$

3. Break even = $146,253 ÷ $\frac{\$264,324}{\$1,619,636}$

4. Break even = $146,253 ÷ 0.1632.

5. Break even = $896,158.

   Our break-even point for the year is $896,158.

6. To find our break-even point for our average month, we divide our break even by 12 ($896,158 ÷ 12 = $74,680).

   Our break-even per month is $74,680.

7. To compute the approximate date that we'll reach our break-even point during the year, we first divide our projected sales volume for the year by 12 to get our projected sales volume for each month ($1,619,636 ÷ 12 = $134,970).

   Our sales volume per month is $134,970.

8. Then we divide our break-even point for the year by our sales volume per month to find our approximate break-even date ($896,158 ÷ $134,970 = 6.639).

   Our approximate break-even date is 6.639, or about June 19.

Answer G: Our approximate break-even date is June 19.

If you worked this problem out, it shows that you have a good grasp of the materials in this book. Congratulations! In my opinion you can now rate yourself in that small group of contractors who will survive and make a success of their business. But don't stop now. There's more!

Problems to Solve

Problem 3: Correcting a Disaster

We have a new-home construction business that's been doing very well — up until about a year ago. Last January, because business was a little slow, someone got the "bright" idea of branching out into remodeling. Several of the families that we'd built homes for in the past had asked us to do some room additions for them. Even though we'd been advised not to mix these two types of construction, it was just too tempting. We caved in and took on two room additions with baths. We promptly lost our assets to the tune of $11,545. Bummer!

In addition to losing money on these two remodeling jobs, we made some unbudgeted purchases, and under stress, got into some disagreements with a couple of customers. We also lost focus on what we should be doing to correct these problems, and instead of increasing our sales, we succumbed to flattery and became sidetracked doing a huge amount of volunteer work at our local association. The result was one very bad year!

We're about to start our eighth year in business, and we've made a joint pact to stay focused and do better.

Based on the last two years' sales, we've made the following projections for the upcoming year:

- We can build six new homes (two each of our most popular plans).

    Plan A, which has projected job costs of $149,880
    Plan B, which has projected job costs of $163,396
    Plan C, which has projected job costs of $185,760

- Our projected overhead is 9.24 percent, or $100,043.

- Our profit goal is 8 percent.

Here is some additional information that we have to consider:

- After a careful review of last year, we realize that we need to recover the $11,545 we lost on the two remodeling jobs we did.

- We've been looking at a set of plans for a commercial remodeling job. Using our new-home square footage price, our preliminary figures show a sales price of $469,000. Another temptation!

- We're thinking about hiring an office manager at a salary of $3,000 per month, which will cost us approximately $3,780 per month.

- We're considering the purchase of a two-year-old one-ton flatbed delivery truck for $19,699, which, including financing and other charges, will cost us $22,654.

Questions:

A. What markup should we use for the commercial remodeling job?

B. What's the correct markup for our company in the coming period, taking into consideration the $11,545 loss from last year?

C. What's the correct selling price for each of the six new homes?

D. If we decide to hire the office manager, how will it affect our markup and sales?

E. Should we buy the truck? (Show the calculations to justify your answer. Don't consider the new office manager in your calculations.)

Problems to Solve

Answers to Problem 3: Correcting a Disaster

A. What markup should we use for the commercial remodeling job?

Answer A: None! We aren't going to pursue the commercial remodeling job, so we won't need to figure a markup for it. Tempting though it is, bitter experience has taught us that mixing new-home construction and remodeling is a recipe for disaster. We'll just tell the customer that we have to pass on this one — our dog, Scrapile, ate the plans.

B. What's the correct markup for our company in the coming period?

1. Let's start by figuring our job costs for the six homes we're planning to build this year.

    Plan A: 2 × $149,880 = $299,760
    Plan B: 2 × $163,396 = $326,792
    Plan C: 2 × $185,760 = $372,520
    Job costs = $998,072

2. Now let's figure out our projected overhead for the year, including the $11,545 loss:

    $100,043 + $11,545 = $111,588
    New overhead is $111,588.

3. Since we know that our overhead is 9.24 percent of our projected sales volume, we can calculate our sales volume by dividing our total overhead by 0.0924.

    $111,588 ÷ 0.0924 = $1,207,662
    Sales volume is $1,207,662.

4. *Sales Volume ÷ Job Costs = Markup*

    $1,207,662 ÷ $998,072 = 1.2099, or 1.21

Answer B: Our markup for the coming year should be 1.21.

C. What's the correct selling price for each of the six new homes?

Job Costs × Markup = Sales Price

Plan A: $149,880 × 1.21 = $181,355 each
Plan B: $163,396 × 1.21 = $197,709 each
Plan C: $185,760 × 1.21 = $224,770 each
Total $603,834 (× 2 = $1,207,668)

Answer C: Plan A: $181,355
Plan B: $197,709
Plan C: $224,770

Total sales for all six new homes: $1,207,668.

D. If we hire the office manager, how will it affect our markup and sales?

There are two approaches we can use to solve this problem. The first is to leave our projected sales as is, and change our markup. The second is to increase our total projected sales for the year, and leave our markup at 1.21. Let's look at what happens if we change our markup.

1. First we need to calculate our new overhead:

Overhead before new office manager	= $111,588
Total cost of the office manager for the year ($3,780 × 12)	= $45,360
New overhead with office manager	= $156,948

2. Now let's figure our profit:

    8% × $1,207,668 = $96,613

3. And now we need to find our new job costs:

    Sales − Overhead − Profit = Job Costs
    $1,207,668 − $156,948 − $96,613 = $954,107
    New job costs are $954,107.

4. *Total Sales ÷ Job Costs = Markup*
    $1,207,668 ÷ $954,107 = 1.2657, rounded up to 1.27

A 0.06 increase in our markup would probably not be an obstacle to sales. The problem with taking this approach is that it reduces the amount of money we'll have to build our jobs. Let's look at the numbers again.

If we increase our overhead by $45,360, we reduce the money available to cover job costs by $43,965 ($998,072 − $954,107 = $43,965). Our new markup will increase our sales by $4,048 (1.27 × $954,107 = $1,211,716 − $1,207,668 = $4,048), but that still leaves us with a $39,917 shortfall. We would have to immediately become more efficient. It would take a 3.29 percent improvement in our efficiency ($39,917 ÷ $1,211,716 = 0.0329) to run our operation profitably on our new budget. That means a lot of belt-tightening and careful management. Could we do it? It might be doable for a short-term expense, but very hard for an ongoing expense like an office manager. What about the alternative approach?

The second approach to this problem is to increase our projected sales to cover the cost of the new office manager. We can use the same formula that we used earlier to include the cost of an unbudgeted expense. The math and the process are exactly the same.

1. Let's calculate our new overhead:

Our existing overhead	= $111,588
The annual cost for the new office manager	= $45,360
Our total new projected overhead expense	= $156,948

Problems to Solve

2. We can use the following formula to find the total annual sales needed to cover the expense of the new office manager:

Overhead ÷ Overhead Percentage = Sales Volume
$156,948 ÷ 9.24\% (0.0924) = \$1,698,571$

We need $490,903 in additional sales this year ($1,698,571 − $1,207,668 = $490,903) to cover our new overhead expense. That means we'll have to build two more Plan C homes with some upgrades (2 × $224,770 = $449,540 + $41,363 in extras) or three more Plan A homes (3 × $181,355 = $544,065) in order to achieve our new sales volume.

We have to decide whether an office manager will free up enough of our time to allow for an almost one-third increase in our business. We'll be increasing our sales and construction from six homes a year to eight or nine. We also must keep in mind that this will be a continuing expense, and that as long we have an office manager we'll have to keep our business at this level.

Answer D: We can raise our markup to 1.27 and increase our efficiency 3.29 percent to cover the expense of an office manager, or we can keep our markup at 1.21 and increase our sales volume to $1,698,571. Since an office manager would be a continuing expense and not just something we could pay off in a year, raising our sales volume would be the better choice. However, now that we know what it will cost us, we'll have to give hiring an office manager some further thought.

Whenever you make a decision like this, you must first evaluate what it will do to your long-term overhead expenses. You must develop a budget for the expense and stick to it, either by adjusting your markup and holding down your job costs, or increasing your sales volume. Many contractors would just go ahead and hire the office manager and not worry about how it would affect their business. Eventually they'd realize that they weren't making enough money to cover their overhead and they'd have to let the office manager go. From that point, they'd be playing catch up on their expenses. That approach to business decisions can only lead one way: down the path to bankruptcy.

E. Should we buy the truck? Show the calculations to justify your answer. (Do not consider the new office manager in your calculations.)

First we need to consider if we really need the truck. We can justify the need if we'll use it 25 days a year or more. Mathematically, we only need to use the truck one day every other week to reach the 25-day requirement (52 ÷ 2 = 26). With our projected sales, we'll easily meet that demand. Now we need to decide if we can work it into our budget.

Let's look at raising our markup during the upcoming year to cover the expense of buying the truck.

1. Our projected sales volume is $1,207,668.

2. Our projected overhead is $111,588.

3. The total cost of new truck is $22,654.

4. Our new projected overhead (with the truck) is $134,242 ($111,588 + $22,654).

5. *Sales − Overhead − Profit = Job Costs*

    $1,207,688 − $134,242 − $96,613 = $976,813

6. *Sales Volume ÷ Job Costs = Markup*

    $1,207,688 ÷ $976,813 = 1.236 or 1.24

    Our adjusted markup for the year would be 1.24.

However, raising our markup without increasing our projected sales will again require some careful management to compensate for the reduction in money available to cover job costs ($998,072 − $976,813 = $21,259 reduction). Let's look at the alternative.

A better approach is to raise our sales volume and leave our markup the same.

7. To calculate our new sales volume:

    New Overhead ÷ Overhead Percentage = New Sales Volume

    $134,242 ÷ 9.24% (0.0924) = $1,452,835

We need an increase in sales volume of $245,167 to recover the cost of the truck in one year ($1,452,835 − $1,207,668 = $245,167). That's one more Plan C home with some extras. However, before we go out and buy that truck, maybe we ought to see if we can get a better deal than $19,699 plus fees and interest!

Answer E: Yes, providing that we will use the truck 25 times during the year and we feel we can sell an additional Plan C home.

Problem 4: All That Glitters Is Not Gold!!!

We have a small company that specializes in roofing work, particularly residential tile roofing. Our four-man crew is the best and fastest in the area, and we can do most homes in two to three days. We do most of our work for general contractors and our average price to them is about $9,900. We estimate that we'll do 225 homes this year, and our overhead for the last three years has been 22.50 percent. We consistently make a 9 percent net profit.

A new general contractor just arrived in town, and he's getting ready to build some very nice homes. We did an estimate for him and came up with the following price to roof each of the models he intends to build:

      Plan A: $14,655

      Plan B: $14,985

      Plan C: $15,235

      Plan D: $15,470

He's planning to build 144 of these homes in his new subdivision over the next two years. He'll turn and flop the plans, ending up with several different versions of these homes, but basically he's going to build each of the four models an equal number of times.

This contractor has seen our work and after a bit of negotiating, has made us the following offer. He'll give us an exclusive contract to roof all the homes, and he'll guarantee us that the homes will be ready at the rate of one every five calendar days. If the homes aren't ready, he'll write us a check for $150 for each day we have to wait before we can start. What kind of idiot would turn his nose up at that offer?

We've checked him through a number of different suppliers and other sources, as well as his bank, and he appears to have both the savvy and the bucks to make this project work. His track record to date for paying his subs has been good. The job looks good.

We discovered, however, that at least a third of the subs who have worked with him over the last three years have gone broke. Hmm.

In our last meeting he "threw in" a couple more requirements that he says he "always asks of the specialty contractors who work on his homes." They are: 1) We must guarantee him that the quote we've given him for each of the four models will remain the same throughout the duration of the project (all 144 homes). 2) We must give him a 20 percent discount on every fourth home.

Over the last three years our material prices have been going up an average of 5 percent across the board every six months. Our labor rates have been going up at the same rate. Our labor and materials expense make up 90 percent of our job costs.

Question:

A. Should we sign the contract? (Show the math to justify your answer.)

Problems to Solve

Answer to Problem 4: All That Glitters Is Not Gold!!!

A. Should we sign the contract? (Show the math to justify your answer.)

There are a number of different ways to calculate the math for this problem and arrive at numbers that will indicate whether we should sign on to work with this guy or pass up the job. We can do the math and incorporate our other work, or focus strictly on this project. It's probably better to stay focused on this project alone. That way we won't be tempted to make adjustments in the sales price of our other work to compensate for lack of profits on this project, if it should it turn out that way. However you choose to do the math, we should both arrive at the same conclusion, assuming we're in business to make a profit.

1. First, let's calculate the total volume of work we're considering.

   Plan A: $14,655 × 36 = $527,580
   Plan B: $14,985 × 36 = $539,460
   Plan C: $15,235 × 36 = $548,460
   Plan D: $15,470 × 36 = $556,920
   Total sales volume = $2,172,420

2. Our overhead expense for this project would be then be $488,795 ($2,172,420 × 22.50% = $488,795).

3. Our 9 percent profit for this project, before deductions, would be $195,518 ($2,172,420 × 9% = $195,518).

4. *Job Costs = Sales Volume − Overhead − Profit*

   Job costs = $2,172,420 − $488,795 − $195,518
   Job costs = $1,488,107

5. Our labor and material expense would be 90 percent of our job costs:

   $1,488,107 × 90% = $1,339,296
   Labor and material expense = $1,339,296

6. Now let's look at the requirement that we deduct 20 percent off the price of every fourth roof that we do. We know that we'll roof an equal number of each of the four models that he's going to build, so we can just divide our total volume by 4 and deduct 20 percent from that to find our total deduction. (We can assume that the deduction will be taken evenly. A well-written contract should prevent the deduction from being taken off only the higher-priced jobs.)

   Total sales: $2,172,420
   $2,172,420 ÷ 4 = $543,105
   $543,105 × 20% (discount) = $108,621
   Total discount = $108,621

244   *Markup & Profit: A Contractor's Guide*

7. We now need to adjust our sales price to reflect the discount:

   $2,172,420 − $108,621 = $2,063,799

   Adjusted sales price = $2,063,799

8. Now let's look at the issue of the 5 percent increase in material and labor expenses that we're anticipating for every six months we'll be on this job. We can figure that the job will take us approximately two years to complete if we're guaranteed one roof job every five days.

   5 × 144 = 720 days ÷ 365 = 1.97 or 2 years

   Since our labor and materials are 90 percent of our job costs, our total labor and material costs will be $1,339,296 ($1,488,107 × 90%). We can divide that by 4 to get the first six months expense:

   $1,339,296 ÷ 4 = $334,824.

   Now we can add the labor and material increase for each six-month period.

   1st six months:
   $334,824 (no increase expected)   = $334,824

   2nd six months:
   $334,824 + $16,741 (5% × $334,825)   = $351,565

   3rd six months:
   $351,565 + $17,578 (5% × $351,565)   = $369,143

   4th six months:
   $369,143 + $18,457 (5% × $369,143)   = $387,600

   Adding up the four six-month figures,
   our total is: = $1,443,132

   We'll assume that the other 10 percent of our costs, $148,810 ($1,488,107 × 10% = $148,810), will remain constant for the duration of the job. Let's add the 10 percent back into our new labor and materials figure to get our adjusted job cost:

   $1,443,132
   + $148,811
   $1,591,943

   Our adjusted job costs are $1,591,943.

9. Now we add our adjusted job costs to our overhead, and deduct it from our adjusted sales price to determine the profit remaining:

Adjusted job costs	$1,591,943
Overhead	+ $488,795
	$2,080,738
Adjusted sales price	$2,063,799
Less job cost and overhead	− $2,080,738
Projected profit / <loss>	<$16,939>

Problems to Solve

We would lose $16,939 of the money needed to pay overhead costs, plus all our projected profit, by guaranteeing the price of this project over the two-year period.

Now we're getting into the philosophy of the business decision-making process. Obviously, we shouldn't take a job knowing we'll lose money. That's a no-brainer. A third of the other contractors that have worked with this guy have done just that, and they're out of business and gone.

However, this project is guaranteed work for two years. That's always nice — as long as it's profitable. To just say No may be a little hasty. Our business is on sound financial footing, and we have other projected sales of $2,227,500 with a 9 percent profit margin. It may be that our costs wouldn't go up that much and we could make a profit. And if our costs did go up, we could probably absorb the $16,939 loss and keep on going. That's the up side. The down side is that an increase in production of this size ($2,227,500 + $2,172,420 = $4,399,920) will also require us to almost double our crew. It'll take time to get the new crew up and working the way we want, and *our overhead will go up* as we add the new people and expenses generated by the additional volume of work. Looking at this proposal from all angles, it isn't a good move for us under the proposed conditions.

We have two choices here. We can walk away from this job, knowing that the contract as proposed won't be profitable. *Or, we can go back to the contractor and negotiate a better deal.*

One approach we could use in negotiations is to say that we'll agree to the contract if he'll allow us to pass along actual material and labor increases during the course of the project, not to exceed 5 percent in any given six-month period. In return, we could guarantee him that we'll start all the new roofs on schedule, but not require him to pay the $150 late fee on any house that he delays, up to, but not exceeding, five days.

As long as we can negotiate a win-win contract, we may still get the contract at a price that's profitable for all of us.

Answer A: Yes, we should sign the contract *if, and only if*, we can negotiate a change in the job requirements that would ensure our business a profit. No, if the contractor refuses to change his requirements.

Even though this is a hypothetical problem, it's very close to what will occur in your real business life. You must evaluate every job offer in this way. If you stay in business long enough, a situation like this will arise, and you'll have to deal with it. If you arrived at the same numbers we did when you worked out this problem, your answer to this offer should have been very similar to the approach we took. If it was, you understand what this book is all about, and you'll do well.

Problem 5: Evaluating a Quarterly Review

You're a contractor and the owner of a one-man remodeling operation called Hammer and Nail Remodeling. You've had a fairly successful couple of years, but you've noticed that business is showing some evidence of decline this year. The Quarterly Overhead Review on page 248 shows your company's expenses during the first six months of this year. You completed $68,600 in work during the first quarter and $77,950 in work during the last quarter. The rise in sales from the first to second quarter of the year is fairly typical for your company. Projected out over the year, you think that you'll do about $293,000 for the year. Let's do some figuring and see if we can put our finger on why business for your company may be off.

Questions:

A. Using the numbers from the last quarter, what markup should your company be using?

B. Looking at the Quarterly Overhead Review, what steps could you take to reduce your overhead and lower your markup?

Problems to Solve

Quarterly Overhead Review
Hammer and Nail Remodeling

Overhead Item	Low / High Percents	Last Quarter	Percent	This Quarter	Percent
1. Advertising	1.50 to 5.00	$2,023	2.94	$2,338	2.99
2. Sales	5.00 to 8.00	$5,145	7.50	$5,846	7.50
3. Office Expenses					
Staff	3.00 to 7.00	$4,699	6.85	$4,699	6.02
Rent	0.35 to 1.20	$515	0.76	$515	0.66
Office equipment	0.10 to 0.50	$137	0.20	$47	0.06
Telephone	0.20 to 0.75	$425	0.62	$492	0.63
Computer	0.10 to 0.35	$123	0.18	$49	0.07
Office supplies	0.05 to 0.20	$41	0.06	$113	0.14
4. Job Expenses					
Vehicles	0.75 to 3.00	$583	0.85	$619	0.79
Job supervision	4.00 to 6.00	$4,116	6.00	$4,677	6.00
Tools & equipment	0.20 to 0.75	$295	0.43	$74	0.09
Service & callbacks	0.10 to 0.50	$75	0.11	$461	0.60
Mobile telephone	0.05 to 0.30	$144	0.21	$171	0.22
Pagers	0.05 to 0.15	$62	0.09	$62	0.08
5. General Expenses					
Owner's salary	6.00 to 8.00	$4,802	7.00	$5,457	7.00
General insurance	0.25 to 1.50	$878	1.28	$1,013	1.29
O.C.R.A.	1.00 to 4.00	$2,058	3.00	$2,339	3.00
Interest	0.50 to 0.75	0	0.00	0	0.00
Taxes	0.00 to 3.00	$480	0.70	$561	0.71
Bad debts	0.00 to 0.30	0	0.00	0	0.00
Licenses & fees	0.10 to 0.25	$82	0.12	$75	0.10
Accounting fees	0.15 to 0.30	$123	0.18	$290	0.38
Legal fees	0.15 to 0.50	0	0.00	$150	0.20
Education & training	0.15 to 0.30	$69	0.10	$125	0.17
Entertainment	0.10 to 0.20	$110	0.16	$84	0.11
Associations fees	0.10 to 0.20	0	0.00	$45	0.06
Totals	23.50% to 53.00%	$26,985	39.34%	$30,302	38.87%

Quarterly overhead review for Hammer and Nail Remodeling Company

Answers to Problem 5: Evaluating a Quarterly Review

A. Using the last quarter numbers, what markup should your company be using?

1. Projected sales volume for the year is $293,000.

2. Projected overhead expense for the year is $113,889 (38.87% × 293,000 = $113,889).

3. Projected net profit is $23,440 (8% × $293,000).

4. Job costs are $155,671. *Total Sales − Overhead − Profit = Job Costs* ($293,000 − $113,889 − $23,440 = $155,671).

5. Markup is 1.88. *Sales ÷ Job Costs = Markup* ($293,000 ÷ $155,671 = 1.88).

Answer A: The markup for your company should be 1.88.

B. Looking at the Quarterly Overhead Review, what steps could you take to reduce your overhead and lower your markup?

Looking at the numbers for your company, the expenses seem normal, some items a little lower than the low/high percentage figures and some a little higher. Overall, the numbers are in line. However, you think that the markup for your company is too high and you'll soon price yourself out of business if you don't bring it into line. There are several things that you can, and should, do to reduce your markup rather dramatically. Since you're running a small one-man operation, you decide that you should make the following adjustments based on the information from the Quarterly Overhead Review.

First, as the owner, you're doing your own sales and job supervision as well as collecting an owner's salary to run the company. For the last quarter, these three expense items added up to $15,980 and amounted to 20.5 percent of the company's overhead. Granted that's a very nice wage, but it's costing the company too much. If you combine these three expenses, since you're only doing each on a part-time basis, you can reduce that to a total overhead figure of 15 percent, which for the three items isn't out of line. That will reduce your quarterly overhead expense by $4,287, and still give the you an average monthly salary of $3,898. If you want to be more conservative in your financial needs, you could even reduce your salary further, to 12 or 13 percent. But let's see how it works with the 15 percent figure. Using the reduced quarterly overhead percentage of 33.37 percent (38.87 − 5.50 = 33.37), how would that change your markup?

1. New projected overhead for the year would be $97,774 ($293,000 × 33.37% = $97,774).

Markup & Profit: A Contractor's Guide **249**

Problems to Solve

2. Profit would still be $23,440 (8% × $293,000).

3. New job costs would be $171,786. *Volume − Overhead − Profit = Job Costs* ($293,000 − $97,774 − $23,440 = $171,786).

4. New markup would be 1.70. *Sales ÷ Job Costs = Markup* ($293,000 ÷ $171,786 = 1.70).

The next item that you want to look at is your office staff. In a small company like yours, the office staff expense should be closer to the minimum 3 percent. The quarterly review shows that this expense is more than double that percentage. Rather than have a full-time book-keeper on staff, you need to either have your bookkeeping subcontracted out, or bring in a bookkeeper once or twice a week for a few hours. Eight hours a week should be more than sufficient time to complete all the necessary bookkeeping for a company of this size. If you can reduce your office expense down to 3.5 percent, it'll reduce your quarterly expense from $4,699 to $2,728 and bring your overhead percentage down another 2.52 percent to 30.85 (6.02 − 3.5 = 2.52 and 33.37 − 2.52 = 30.85). Let's see what this will do for your markup.

1. New projected overhead for the year would be $90,391 ($293,000 × 30.85% = $90,391).

2. Profit would still be $23,440 (8% × $293,000).

3. New job costs would be $179,169. *Volume − Overhead − Profit = Job Costs* ($293,000 − $90,391 − $23,440 = $179,169).

4. New markup would be 1.64. *Sales ÷ Job Costs = Markup* ($293,000 ÷ $179,169 = 1.64).

You think you might be able to make further reductions in your expenses by using your company vehicles more economically (i.e. reducing the mileage driven), renting rather than buying tools and equipment when you need them, and possibly reducing your O.C.R.A. expense by $1/2$ to 1 percent.

As you can see, with a little more attention to the details of your business, you can reduce your markup from 1.88 to at least 1.64 and possibly more. This will put you in a far better position for sales.

Answer B: You can reduce your overhead by 8 percent (a 5.5 percent reduction in the combined owner's salary expense, job superintendent expense and sales expense, and another 2.5 percent reduction in your office staff expense), thus lowering your markup from 1.88 to 1.64.

With proper planning and expense evaluation you can reduce your markup, and at the same time maintain the income necessary to cover all of your job costs, all of your overhead expenses and still make 8 percent net profit. The important point to remember is that you must *always* do the math and be aware of what your actual expenses are and where they come from.

Problem 6: Boom & Bang Construction, Inc.

Your company has just finished its eleventh year in business, and you've completed nearly 510 jobs. You specialize in high-end residential remodeling. Your company has done some commercial jobs, but it's very seldom that you do that type of work. You have seven full-time employees including yourself, four employees in the field and two in the office. You also have two salespeople who work as independent contractors and get paid a flat 7 percent commission on all sales that meet a minimum 8 percent net profit. As independent contractors, they don't receive company benefits. You use subcontractors for almost everything on the job.

Last year, you did $2,644,945 in good, solid business. Next year you plan on a 10 percent increase in sales. Last year you wanted to make an 8 percent net profit, but actually made only 5.8 percent. Your goal for next year is to meet that 8 percent profit.

The good news is that you're in excellent health, and you're the sole proprietor of your own business. Of course, this came about due to the unfortunate death of your business partner last December. Boom & Bang Construction is known far and wide for quality work at a fair price. You don't believe in recessions, so you've never participated in one. You and your partner made your business what it is, and it's considered a model construction company — thank you very much! Now, here's the rest of the story.

- You dearly love your wife and nine kids, and you may be expecting twins! Your wife told you last night that she's signed you both up for marriage counseling. She thinks there should be something more to life than having and raising children. The two youngest children got low marks on their report cards, and your oldest son received a three-day suspension from school for starting a food fight in the cafeteria. This could be the first indication of poor discipline at home. Your wife thinks you need to spend more time at home with your children and less time working.

- Your best job superintendent, the one who's been with you since day one, came into your office yesterday morning and asked for raise. She says if she doesn't get the raise, she's moving on. She also might be filing a sexual harassment suit. She referred to the new superintendent that your late partner hired, and that she's had to train, as "the octopus!" In the middle of your conversation, the phone rang and you said you'd get back with her.

- The phone call was from your banker. Just a friendly reminder that your payment on the new Mercedes is late! He also wanted to know when he can collect on that dinner for eight that you bet him last week on the eighteenth hole at the country club. That was the really

Problems to Solve

easy two-foot putt you missed, remember? You figure the dinner will cost you at least $300, maybe a lot more if everyone drinks as much as you're afraid they will.

- You were notified last Thursday that the building your office is in has been sold. You have to be out in 45 days. You've been paying $650 a month rent and you know that your rent anywhere else will be at least 25 percent higher. Your best estimate is that it will cost you about $4,500 to move.

- Your partner died suddenly on December 31 last year, leaving several hundred thousand dollars of personal bills, and the only asset that his estate shows is 49 percent ownership in Boom & Bang Construction. Fortunately, you and your partner had a buyout agreement that guaranteed the surviving partner could buy out the other partner's stock in the company for $1. You executed that clause on the day after the wake. At the time of his passing, you were each paid an equal salary.

- The attorney for your late partner's estate called to say that he thinks it wouldn't be wise for you to use money from Boom & Bang Construction, Inc. to pay any more company bills until he's had a chance to audit the corporate books. He wouldn't tell you why, he just mumbled something about illegal agreements and the IRS. He can work the audit into his schedule sometime in the next three to four weeks.

- You had Key Man Insurance, but you don't know if your bookkeeper paid the premiums when they were due this year. You can't find the file for the insurance policies. Your bookkeeper is on vacation on a remote island in the South Pacific and can't be reached. He's been gone for three weeks, and won't be back for another two. You wonder how he can afford such a trip on what he earns.

- You were appointed chairman of your local association's committee to determine what the proper markup for remodeling contractors should be. They want to set a standard for the members. You were thrown out of the first committee meeting, called a heretic and charged with blasphemy. You were told that nobody in the remodeling industry could possibly sell their jobs with a 1.50 markup; customers simply won't pay such exorbitant fees! Tar and feathers were mentioned if you should return.

- A friend who owns an automobile dealership called yesterday morning to tell you that he just got a two-year-old one-ton van with only 14,842 miles on it in on a trade. He only wants $7,500 for it, and you desperately need that van. You didn't plan a new vehicle into your budget. Your banker says he can give you a loan for $221.06 a month for three years, with a 10 percent down payment. Your dealer can arrange a lease at about $241.86 a month, and it would have a $1 buyout after three years.

- You've recently been hit with two lawsuits. It seems that your truck driver has a habit of going into your customers' refrigerators after delivering items into their homes, and helping himself to lunch. One of your customers, Mr. Authause, has taken exception to this, and his attorney has dubbed this the "Fridge-Raider" lawsuit. Your attorney says this will cost you a minimum of $1,500 in legal fees, which she wants up front. The other lawsuit is frivolous, so says your attorney, and you don't need to worry about it . . . yet!

- The insurance agent that handles your Workers' Compensation insurance just called. He said your rates are going way up next year; "sky high" was actually the term he used. You had a brief discussion about why that would be, since your company has an extremely good safety record, and he said he'd look into it. He told you not to worry. He was headed for his favorite watering hole right then, but he'd certainly have the problem settled in a day or two — hopefully before he leaves on his three-week vacation Wednesday. In the meantime, your new premiums are due.

- Your top contract salesperson, Pete, has been having some personal problems lately. You noticed that your late partner approved a couple of draws against future sales, and Pete's account is now $4,500 in the red. He hasn't taken any leads in almost two weeks, and hasn't sold anything in over a month. You've been looking for Pete's contract with the company, but can't find it. That darned bookkeeper; you need to talk with him, and with Pete!

- Your new superintendent (the octopus) just told you that your electric roto-hammer has to have the motor rewound. That will cost you $385.

- Your company picnic is scheduled for two weeks from Saturday, and the food will cost at least $650.

- Your remodeling company has six jobs in the works right now, with a total volume of $256,934 when completed.

- You've decided that since you now have the full burden of running the company, you'll give yourself a raise, your owner's salary plus one quarter of your late partner's salary, starting immediately.

Problems to Solve

The review form on page 255 shows the overhead costs for your company as of December 31 of last year, as best as you can determine without the help of your bookkeeper.

Questions:

A. Using your projected sales figures for next year, what should your markup be after you incorporate all your new expenses?

B. What are your margin and your percent of margin at this point in time?

C. If average production for a well-run company is approximately $170,000 to $180,000 per employee, is your company overstaffed, understaffed or staffed just right?

D. Using your projected sales volume, what effect will your partner's death have on your markup?

E. Should you buy the van, lease the van, or just pass on the van?

F. Your biggest competitor, whose office is just four blocks from your new office, has lowered his markup from 1.63 to 1.55. What should your new markup be?

G. In March, the building you're moving your office into will be up for sale. You could make an offer to buy the building and take possession on April 1st. If you did, your payments would be $3,000 a month for 10 years. What would your new volume sold, built and collected have to be for the remainder of the year in order to make these payments on the building? What would your new markup be? Should you buy the building? (Show the math for all your answers.)

Overhead expenses for Boom & Bang Construction

Overhead Review
Boom & Bang Construction Company

1. Advertising:	$52,899
2. Sales:	$185,146
3. Office Expenses:	
Office staff	$55,544
Rent	$650
Office equipment	$2,645
Computer & peripherals	$5,290
Telephone	$4,629
Office supplies	$1,323
4. Job Expenses:	
Vehicles	$27,772
Job supervision	$158,696
Tools & equipment	$17,192
Service & callbacks	$9,257
Mobile telephone	$2,116
Pagers	$1,058
5. General Expenses:	
Owner's salary (as partner)	$105,798
General insurance	$18,515
Interest expense	$5,290
Bad debts	0
O.C.R.A.	$39,799
Licenses & fees	$5,290
Accounting fees	$5,819
Legal fees	$4,232
Education/training	$4,761
Entertainment	$2,645
Association Fees	$800
Total	$717,166

Problems to Solve

Answers to Problem 6: Boom & Bang Construction, Inc.

First things first. All business owners have their own set of problems. The first thing you need to do is separate the wheat from the chaff. Don't let your personal life confuse what you need to deal with in your business life. They're two different arenas. Though they're both important, only business matters should be considered when you're dealing with your business. Let's take a look at the items you set out on the table and decide which are business expenses and which are personal expenses. (As a business owner, there's a temptation to consider *everything* a business expense. Don't do it. The IRS is on the lookout for this kind of thing, and they have unlimited manpower and a bottomless pocket. You don't.)

- Family matters are family matters, and not to be considered here. Family and/or marriage counseling is a personal expense.

- Before you address the issue of a raise for your lead superintendent, you'd better look into her sexual harassment claim. This isn't something that you can put aside. Once that's settled, and you've evaluated your overhead expenses for the year, you can look into adjusting her salary. If and when you arrive at a new number for her, you can then adjust the job supervision column in your overhead expenses and recompute your markup.

- The Mercedes is your personal car and should be considered your personal expense. It doesn't belong in your business expenses.

- Dinner for eight for your banker is also a personal item, not to be included in the company expenses. Even though your banker is a business associate, the IRS doesn't recognize golfing wagers as a business expense. Pity — if you're a bad golfer.

- Your office move and the increase in rent are valid business expenses.

- Your partner's death and subsequent buyout is a business expense.

- Refer any discussions with your late partner's attorney about illegal agreements with the IRS to your attorney. Stay focused on your business.

- Deal with the issue of Key Man Insurance when your bookkeeper returns.

- Resign the chairmanship of your association committee on markup immediately. If they aren't interested in what you have to say, drop it. Focus on running your own company.

- The van is a valid business expense. We'll consider it in more detail.

- Legal fees for business-related lawsuits (truck driver's lunches) are valid expenses. You need to address this issue immediately.

- Deal with the Workers' Compensation insurance when you get the invoice and you have a firm number to work with. Stay focused on the things you can control.

- Get the issue with Pete resolved immediately before it becomes a bad debt expense. Don't wait for your bookkeeper to come back from vacation. Pete may just need a pep talk and some encouragement to get him back on track.

- The roto-hammer repair is a valid expense.

- The company picnic is also a valid business expense.

- If your salespeople have compiled accurate estimates for the six jobs in the works, and you've checked their math, then these jobs are already part of your overhead job costs. You don't need to deal with this information.

- An increase in your salary is a valid expense.

Now that we've established what are and what aren't valid business expenses, we can plug them into your overhead for next year as shown on page 258. We can also upgrade it to reflect your new sales projections.

A. Using your projected sales figures for next year, what should your markup be after you incorporate all your new expenses?

1. Projected sales volume is $2,909,440.

2. Projected profit at 8 percent is $232,755.

3. Projected overhead at 28 percent is $814,578.

4. *Job Costs = Sales Volume − Overhead − Profit*
   $2,909,440 − $814,578 − $232,755 = $1,862,107
   Projected job cost is $1,862,107.

5. *Markup = Sales Volume ÷ Job Costs*
   $2,909,440 ÷ $1,862,107 = 1.5625 or 1.57

Answer A: Your new markup should be 1.57

B. What are your margin and your percent of margin at this point in time?

1. Margin:
   Overhead + Profit = Margin
   $814,578 + $ 232,755 = $1,047,333

Problems to Solve

Adjusted overhead expenses for Boom & Bang Construction

Overhead Expenses
Boom & Bang Construction Company

1. Advertising:	$58,189
2. Sales:	$203,661
3. Office Expenses:	
Office staff	$61,098
Rent	$813
Office equipment	$2,909
Computer & peripherals	$5,819
Telephone	$5,092
Office supplies	$1,455
4. Job Expenses:	
Vehicles	$30,549
Job supervision	$174,566
Tools & equipment	$18,911
Service & callbacks	$10,183
Mobile telephone	$2,328
Pagers	$1,164
5. General Expenses:	
Owner's salary (+ $\frac{1}{4}$ partner's salary)	$132,248
General insurance	$20,366
Interest expense	$5,819
Bad debts	$2,909
O.C.R.A.	$43,642
Licenses & fees	$5,819
Accounting fees	$6,401
Legal fees	$4,655
Education/training	$5,237
Entertainment	$2,909
Association fees	$800
Total	**$807,542**
6. **New Expenses:***	
Office move	$4,500
Partner buyout	$1
"Fridge-Raider" lawsuit	$1,500
Picnic	$650
Roto-hammer repair	$385
New Total	**$814,578**

*****Note:** The items under heading number 6, *New Expenses*, should ordinarily be included under their specific item headings in listings 1 through 5 above. For illustration purposes only, we've separated them out here under an additional heading.

2. Percent of margin:
   Margin ÷ Sales Volume = Percent of Margin
   $1,047,333 ÷ $2,909,440 = .35998 or 36%

Answer B: Your margin is $1,047,333.
Your percent of margin is 36 percent.

C. If average production for a well-run company is approximately $170,000 to $180,000 per employee, is your company over-staffed, understaffed or staffed just right?

1. *Sales Volume ÷ Staff = Production Per Employee*
   $2,909,440 ÷ 7 = $415,634 per employee

It's obvious from these figures that your company is doing very well as far as production per employee. Your sales volume could easily justify 16 employees ($2,909,440 ÷ $180,000 = 16). Hiring independent contractors for sales is one way that you've kept your staff down and production up. Good for you!

Answer C: Your production per employee rate indicates that your company could easily support more employees, but since you have a successful balance worked out, your staff appears to be just right.

D. Using next year's projected sales volume, what effect will your partner's death have on your markup?

Before partner's death:

1. Sales volume for last year was $2,644,945.
2. Overhead for last year was 27.12 percent, or $717,166.
3. Profit last year was 5.8 percent, or $153,407.
4. *Job Costs = Sales Volume − Overhead − Profit*
   $2,644,945 − $717,166 − $153,407 = $1,774,372
5. *Markup = Sales Volume ÷ Job Costs*
   $2,644,945 ÷ $1,774,372 = 1.49
6. Markup last year was 1.49.

After death of partner:

1. Projected sales volume for this year is $2,909,440.
2. Projected overhead for this year at 28 percent is $814,578.
3. Projected profit for this year at 8 percent is $232,755.
4. *Job Costs = Sales Volume − Overhead − Profit*
   $2,909,440 - $814,578 - $232,755 = $1,862,107

Problems to Solve

5. *Markup = Sales volume ÷ Job cost*
   $2,909,440 ÷ $1,862,107 = 1.5624

6. Markup for this year is 1.57.

Answer D: The death of your partner, as well as your other new expenses, will cause your markup to increase from 1.49 to 1.57.

E. Should you buy the van, lease the van, or just pass on the van?

Let's consider this questions from a few different angles.

1. Your company overhead is a little too high right now and you need to make some adjustments to bring it down. How do you know that? Plain and simple, your markup went up 8 points this year already. While not a disaster, it should raise some immediate concerns and be a warning indicator that you need to slow down on expenditures.

2. You may argue that, in light of your really good employee production rate, the company is very stable and one purchase will not take it down. However, consider the following problems that may be pending:

   (a.) What happens if your late partner's estate sues and wins a judgment for $500,000 or more? Do you have insurance that would cover this type of expense?

   (b.) You can't even *find* your Workers' Comp file, or Pete's contract! What would you do if your bookkeeper didn't come back from the South Pacific? What if he's taken off with company funds? You need to get your office better organized, find out just where your files are and learn how to check up on your company's financial health on your own.

   (c.) Suppose you can't appease your job superintendent and you have to deal with a sexual harassment lawsuit? Do you have insurance to cover that?

   (d.) What if your late partner got your company into some problems with the IRS?

   (e.) Have you considered the consequences resulting from the "Fridge-Raider" lawsuit? Aside from the costs directly involved, there could be negative publicity that may cost you some business. Most people don't like tradesmen rummaging through their kitchen. You'll need to invest some additional funds into advertising and PR to reinforce your reputation once this nonsense hits the media (and you can bet it will). By the way, you did fire that dude, didn't you? If not, why not?

Under your present circumstances, buying or leasing the van isn't a good business decision. Next year things will be more settled, and another van will come along.

Answer E: At this time, you should pass on the van.

F. Your biggest competitor, whose office is just four blocks from your new office, has lowered his markup from 1.63 to 1.55. What should your new markup be?

Answer F: 1.57.

(If there is even the slightest doubt about your answer to this question, go back and reread Chapter 3.)

G. In March, the building you're moving your office into will be up for sale. You could make an offer to buy the building and take possession on April 1. If you did, your payments would be $3,000 a month for 10 years. What would your new volume sold, built and collected have to be for the remainder of the year in order to make these payments on the building? What would your new markup be? Should you buy the building?

First let's look at what your new volume sold, built and collected will have to be if you make this purchase.

1. Projected overhead for the coming year is $814,578.

2. Projected overhead per month ($814,578 ÷ 12) is $67,882.

3. New building expense for the rest of the year (9 × $3,000) is $27,000.

4. Projected overhead, April to December:
   $67,882 × 9 = $610,938 + $27,000 = $637,938

5. Using the following formula to project the needed increase in sales:
   New Overhead ÷ Percent of Existing Overhead = New Sales Volume
   $637,938 ÷ 28% = $2,278,350

The sales volume required to pay for a new building, from April to December, is $2,278,350. That would increase your overhead to $70,882 per month for the remaining nine months of the year. And, since this purchase is over a 10-year period, your high overhead and increased sales volume would have to continue until the building is paid off.

Now let's look at what your new markup would be.

1. Your new profit would be $182,268 (8% of $2,278,350).

2. *Sales Volume − Overhead − Profit = Job Costs*
   $2,278,350 − $637,938 − $182,268 = $1,458,144

Problems to Solve

3. *Sales Volume ÷ Job Costs = Markup*
$$\$2{,}278{,}350 \div \$1{,}458{,}144 = 1.563 \text{ or } 1.57$$

Your markup would remain at 1.57.

Should you buy the building? As we discussed when we were dealing with the question of purchasing the van, now is not the time to take on additional expenses. You need to "clean up your act" and get all of your pending problems settled before you take on anything else. However, you could possibly negotiate an option to have the first right of refusal for purchase of the building for one year. Then, when things have calmed down a bit, when all the legal issues have gone away, and you've reviewed and reduced some of your overhead expenditures, you could follow up on the purchase of the new office building. This "cleaning up" process will normally take from six months to a year or more, depending on how each of these problems gets worked out.

In the meantime, do your homework on the building. Get the information you need to determine what it will cost initially to buy the building and what it will cost for the long term to maintain your purchase. Then consider what you can do with the building in terms of a showroom, office space, equipment storage, and so on. What's the electric power, water and sewage situation to the building? Find out what the city or county is doing around the area. Are they planning a new freeway right through the middle of the property? Check with the city or county department that projects urban growth. Is the building in the right spot to take advantage of increased population growth in your area? Could you lease the building out to a retail outlet, move your company to another building, and make enough money from the lease to pay for the building and the rent or mortgage on another building in a location that's good, but perhaps not quite as good? Get your game plan together while you solve your other problems. Give yourself some room to gather the information you need to make some good decisions for Boom & Bang Construction, Inc. and its future growth. It deserves your best effort.

Answer G: Your new volume sold, built and collected would need to be $2,278,350. Your markup would remain at 1.57.

You shouldn't buy the building at this time, but you could try to negotiate an option for first right of refusal for purchase for one year so you can pursue the purchase when you get your business affairs in order.

Congratulations!

You've completed all the problems and you're ready to put this information to work for you. You now have the tools you need for success. I hope I've made you think and helped you develop the skills you'll need to manage your own business successfully. Give yourself a pat on the back for a job well done!

Appendix A

Blank Forms, on Paper and Disk

The following pages contain blank copies of the forms and checklists featured in this book, plus several additional forms that should help you keep your finances, your office and your jobs in order. These additional forms come from *Construction Forms & Contracts*, also by Craftsman. See the order form in the back of this book.

You have the publisher's permission to copy any or all of these forms to use in your business. Paste or stamp your own letterhead or logo in the space at the top, then use your copy machine.

If you use a computer in your office, simply load the files from the CD-ROM included inside the back cover of this book, customize them to suit your needs, fill them out, and print.

Installing the CD if You Have Windows

You can use any form right off the CD if you want. Instructions for this are on the next page. But it's better to install the forms on the hard drive of your computer. When installation is done, there's no need to keep the CD in the CD drive any longer.

To install the forms onto your hard drive, turn your computer on and start Windows. Put the *Markup & Profit* disk in your CD drive (such as D:).

In **Windows 3.1 or 3.11**
go to the Program Manager.

1. Click on **File**
2. Click on **Run . . .**
3. Type D:\SETUP
4. Press **Enter**

In **Windows 95** or higher

1. Click on **Start**
2. Click on **Run . . .**
3. Type D:\SETUP
4. Press **Enter**

Then follow the instructions on the screen. We recommend accepting the default folder (C:\Markup). Select the word processing program of your choice. The forms will be installed to your hard drive in this format. Click on **Next** to continue. When installation is complete, click on **Finish**.

Uninstalling for Windows 95, 98 and NT

Click on **Start**, then point to **Programs**, **Markup & Profit**, then click on **Uninstall**.

Shortcuts for Windows 95, 98 and NT

Once *Markup & Profit* has been installed, there's an easy way to open the form of your choice:

1. Click on **Start**
2. Click on **Programs**
3. Click on **Markup & Profit**
4. Click on **Forms**
5. You'll see the names of all forms installed
6. Double click on any form name to start the associated program with the form already open.

To Use the Forms Right Off the CD

1. Start your word processing program
2. Click on **File**
3. Click on **Open**
4. Switch to your CD drive (usually D:)
5. Double click on the folder for the word processing program you're using (Word2, Word97 or WordPerfect)
6. Double click on the name of any form you want to see
7. If you make changes, be sure to save the revised form to your C: drive.

Installation on a Macintosh

All the construction forms are on the CD-ROM in the folder WordMac. You can use these forms right off the CD-ROM (without installing). First, make sure the *Markup & Profit* disk is in the CD drive. Start Word. Open any form on the disk in the WordMac folder.

We recommend installing these forms to a new folder on your hard drive. Here's how:

1. From the startup screen, click on your hard drive icon
2. You'll see a list of folders currently on the hard drive
3. Click on **File**
4. Click on **New Folder**
5. A new untitled folder will appear in a black box on the folder list

Appendix A: Blank Forms, on Paper and Disk

6. Type *Markup* over untitled folder. This is where you'll store the forms on your hard drive.
7. Press **Enter** and the Markup folder will move into alphabetical order on the folder list
8. Double click on the Markup icon to open the Markup folder
9. Drag and drop the WordMac icon in the Markup folder.

Opening Forms in Word for the Mac

1. Start Microsoft Word
2. Click on **File**
3. Click on **Open**
4. At the Desktop, double click on the hard drive icon
5. Scroll down until you find the Markup folder
6. Double click on **Markup** and then double click on **WordMac**
7. Scroll down the list until you find the form you want to open
8. Double click on the form name
9. Click on **File** and **Print Preview** to see if the form is centered. If not, close Print Preview. Drag the left or right margin until the form centers.
10. When printing, you may see a warning about margins. If so, click on **Fix** and the margins should reset correctly for your printer.

Directory of Blank Forms

Description	Filename	Description	Filename
Quotation Request	QUOTE	Delay Notice	DELAYNOT
Job Estimate	JOBEST	Potential Backcharge Notice	POTBACKNO
Estimate Checklist	ESTCHECK	Final Completion List	JOBCOMPL
Specialty Selection Sheet	SPECSEL	Project Closeout Checklist	PROJCLOS
Project Management Contract	PROJMAN	Monthly Sales Report	SALESREP
Layout Sheet	LAYOUTSH	Quarterly Overhead Review New Home Construction	QOVHDNEW
New Job Checklist	NEWJOB	Remodeling	QOVHDREM
Job Schedule	SCHEDUL	Specialty	QOVHDSPC
Change Order	CHANGE	Overhead Expense Chart	OVERHEAD
Subcontractor Change Order	SUBCONTC	Profit & Loss Statement	PROF&LOS
		Break-Even Point Worksheet	BREAK

Quotation Request

Date:	Request taken by:
Customer name:	

Job address:

Mailing address:

Home phone: ( )
Home fax: ( )
His work: ( )
His fax: ( )
Her work: ( )
Her fax: ( )

Best time to call: _____ AM/PM
His pager: ( )
His mobile: ( )
Her pager: ( )
Her mobile: ( )
Referred by:

Job

Major remodel	Room Addition	Kitchen	Bath
Sunroom	Dormer	Basement	Siding
Deck / Exterior	Handyman	New home	Commercial

_____
_____
_____
_____

Other information:
_____
_____

Appointment time: _____ AM/PM Day: _____ Date: ____ / ____ / ____

Directions to job site:
_____
_____
_____
_____
_____

Referred to:

QUOTE

Job Estimate

Estimated by:	Date:
Customer:	
Address of Job:	

#	Item	Labor	Materials	Subcontractor	Other
1	General conditions				
2	Demolition / tear out				
3	Excavation				
4	Concrete				
5	Masonry				
6	Framing				
7	Roofing				
8	Siding				
9	Windows				
10	Doors				
11	Sheet metal				
12	Plumbing				
13	Electrical				
14	H.V.A.C.				
15	Insulation / weatherstripping				
16	Drywall / plaster				
17	Ceiling tile				
18	Cabinets				
19	Surfacing				
20	Tile				
21	Floor covering				
22	Kitchen & bath accessories				
23	Awning & patio				
24	Finish carpentry				
25	Hardware & metalwork				
26	Paneling & fence				
27	Light fixtures				
28	Paint & decor				
29	Debris removal				
30	Miscellaneous				
	Estimated totals				

Job total	
O & P	
Subtotal	
State sales tax	
Quote	

Estimate Checklist

Task description	Materials				Labor			Sub bids/ Equip. rental	Task total	Notes/ allowances
	Unit	Amt.	Cost	Total	Unit	Amt.	Cost	Total			
Planning & initialization											
Building permits											
Construction heating											
Inspection fees											
Office trailer setup											
Plans											
Safety barriers											
Sewer connection fee											
Temporary power											
Water meter fees											
Other											
Site work											
Backfill & grading											
Basement excavation											
Earth imported/export											
Footings excavation											
Off site improvements											
Sewer lateral/waterline											
Site preparation											
Soil treatment											
Other											
Concrete											
Flatwork											
Footings											
Foundation											
Additives											
Area wells											
Bolts											
Chamfer strips											
Cold joint											

Estimate Checklist (Continued)

Task description	Materials				Labor				Sub bids/ Equip. rental	Task total	Notes/ allowances
	Unit	Amt.	Cost	Total	Unit	Amt.	Cost	Total			
Columns											
Drayage											
Expansion joint											
Form oil & cleaner											
Forms & whalers											
Forms, stakes, braces											
Poly/visqueen											
Rebar-1/2" @ ft.											
Rebar-3/8" @ ft.											
Rebar-5/8" @ ft.											
Rebar-7/8" @ ft.											
Rental equipment											
Straw/hay											
Ties & tie wire											
Waterproofing foundation											
Window & door bucks											
Wire mesh											
Other											
Rough lumber											
Studs 2 x 4											
Studs 2 x 6											
Trusses											
Fir 2 x 8 x 8											
Fir 2 x 8 x10											
Fir 2 x 8 x12											
Fir 2 x 8 x14											
Fir 2 x10 x10											
Fir 2 x10 x12											
Fir 2 x10 x14											
Fir 2 x12 x10											
Fir 2 x12 x12											
Fir 2 x12 x14											
Fir 2 x12 x___											

Estimate Checklist (Continued)

Task description	Materials				Labor			Sub bids/ Equip. rental	Task total	Notes/ allowances
	Unit	Amt.	Cost	Total	Unit	Amt.	Cost	Total			
1 x 4 #2 Grade											
1 x 6 #2 Grade											
1 x 8 #2 Grade											
1 x 10 #2 Grade											
Post-4 x 4"											
Post-6 x 6"											
Glulam											
Plywood 3/8"											
Plywood 1/2"											
Plywood 5/8"											
Plywood 3/4"											
Plywood T & G 5/8"											
Plywood T & G 3/4"											
Fascia											
Soffit											
Frieze											
Caulking & sealants											
Builders adhesive											
Scaffolding											
Equipment rental											
Nails 16d box											
Nails 16d nail gun											
Nails 8d box											
Nails 8d nail gun											
Nails 16d duplex											
Nails 8d duplex											
Nails-finish (rough only)											
Bolts, washers, nuts											
Framing connectors											
Saddles											
Staples, air gun											
Other											
Roofing											
Built-up											

Estimate Checklist (Continued)

Task description	Materials				Labor				Sub bids/ Equip. rental	Task total	Notes/ allowances
	Unit	Amt.	Cost	Total	Unit	Amt.	Cost	Total			
Tile											
Shake											
Cedar											
Drying-in											
Felt-#15											
Felt-#30											
Ice & water shield											
Flashing & counter F.											
Drip edge											
Valley tin											
Roof jacks & vents											
Galv. roof nails-3/4"											
Galv. roof nails-1-1/2"											
Simplex nails											
Roofing cement											
Other											
Insulation											
Walls											
Ceiling											
Miscellaneous insulation											
Radiant barrier											
Vapor barrier											
Drywall											
Corner bead											
Drytite nails- @ lb.											
Hot mud											
J-Metal											
Knife blades											
Masking supplies											
Perfa-tape											
Quick-tape											
Topping-boxes											
Other											

ESTCHECK

Estimate Checklist (Continued)

Task description	Materials				Labor				Sub bids/ Equip. rental	Task total	Notes/ allowances
	Unit	Amt.	Cost	Total	Unit	Amt.	Cost	Total			
Finish carpentry											
Base molding											
Chair rail											
Crown molding											
Doors											
Door casing											
Front door											
3/0 exterior											
2/8 exterior											
Fire rated/self-closing											
3/0 Hollow Core interior											
2/8 H. C. interior											
2/6 H. C. interior											
2/4 H. C. interior											
2/0 H. C. interior											
Bi-fold 7'											
Bi-fold 6'											
Bi-fold 5'											
Bi-fold 4'											
By-pass 7'											
By-pass 6'											
By-pass 5'											
By-pass 4'											
Pocket 3/0											
Pocket 2/8											
Pocket 2/6											
Pocket 2/4											
Specialty doors											
Garage door											
Shelving 1 x12											
Shelving 1 x16											
1 x 4" pine											
1 x 6" pine											
1 x12" pine											

Estimate Checklist (Continued)

Task description	Materials			Labor			Sub bids/ Equip. rental	Task total	Notes/ allowances
	Unit	Amt.	Cost	Total	Unit	Amt.	Cost	Total	
3/4" Ply/pressed wood										
Shims										
Keyed locks										
Deadbolts										
Privacy locks										
Passage locks										
Other locks/latches										
Pulls & catches										
Door bumpers/stops										
Towel bars/rings										
Paper holders										
Soap dishes										
Closet rods										
Robe hooks										
Other bathroom hardw.										
Door viewer										
House numbers										
Mail box										
Garage door operator										
Thresholds										
Weatherstripping										
Screen doors										
Other										
Wall coverings										
Acoustical ceiling tile										
Ceramic tile										
Masking supplies										
Paint rollers/brushes										
Paint-alcohol sealer										
Paint-latex										
Paint-oil										
Paint-PVA										
Paneling (sheets)										

ESTCHECK

Estimate Checklist (Continued)

Task description	Materials				Labor				Sub bids/ Equip. rental	Task total	Notes/ allowances
	Unit	Amt.	Cost	Total	Unit	Amt.	Cost	Total			
Sandpaper, caulking, etc.											
Stain & varnish											
Textured paints											
Wainscot											
Wallpaper tools/glue											
Wallpaper/border (rolls)											
Other											
Floor covering											
Area rugs											
Asphalt/rubber tile											
Base (rubber)											
Carpet glue											
Carpet tool rental											
Carpets											
Ceramic tile											
Concrete sealer											
Cove											
Hardwoods/parquet											
Linoleum											
Linoleum glue											
Metal trim											
Nails & glue											
Other exterior carpets											
Other floor paint											
Padding											
Quarry tile											
Rock											
Roller & blades											
Seam filler											
Seam tape											
Staples											
Synthetic turf											
Tack strips											
Other											

Estimate Checklist (Continued)

Task description	Materials			Labor			Sub bids/ Equip. rental	Task total	Notes/ allowances
	Unit	Amt.	Cost	Total	Unit	Amt.	Cost	Total			
Glazing											
Windows											
Patio sliding door											
Stained glass											
Sandblast/etched glass											
Other glazing											
Specialty glass											
Mirrors											
Medicine cabinets											
Skylights											
Other											
Masonry											
Fireplace											
Fireplace face											
Veneer											
Masonry sidewalks/paths											
Fence											
Other											
Plumbing											
Rough											
Finish											
Jacuzzi - spa											
Shower stall											
Tub - enclosure											
Extra water heaters											
Fixture allowance											
Water softener											
Sprinkling sys./rough											
Other											
HVAC											
Furnace & distribution											
Attic vents											

Estimate Checklist (Continued)

Task description	Materials			Labor			Sub bids/ Equip. rental	Task total	Notes/ allowances
	Unit	Amt.	Cost	Total	Unit	Amt.	Cost	Total			

Task description	Unit	Amt.	Cost	Total	Unit	Amt.	Cost	Total	Sub bids/ Equip. rental	Task total	Notes/ allowances
Bathroom vents											
Kitchen vent											
Other vents											
Evaporative cooler											
Air conditioning											
Special controls											
Area heater											
Wood-burning stove											
Fireplace - gas log											
Fireplace blower											
BBQ grill											
Humidifier											
Electronic air purifier											
Other											
Electrical											
Rough											
Finish											
Light fixture allowance											
Landscape light(s)											
Connect HVAC, etc.											
Doorbell/chime											
Security system											
TV/ stereo sys/antenna											
Intercom system											
Computer system											
Smoke/fire alarm sys											
Telephone installation											
Other											
Cabinets											
Kitchen cabinets											
Bath cabinets											
Linen closets											
Bookshelves											

Estimate Checklist (Continued)

Task description	Materials				Labor				Sub bids/ Equip. rental	Task total	Notes/ allowances
	Unit	Amt.	Cost	Total	Unit	Amt.	Cost	Total			
Pantry											
Clothes chute											
Garage - cabinets											
Garage - workbench											
Other											
Hard surfaces/counters											
Ceramic tile											
Corian											
Extra materials											
Plastic laminate bid											
Synthetic marble											
Other											
Appliances											
Countertop unit											
Dishwasher											
Disposal											
Freezer											
Microwave oven											
Range											
Range hood											
Refrigerator											
Separate oven(s)											
Trash compactor											
Under cabinet units											
Other											
Exterior wall coverings											
Gutter & downspout											
Rock											
Siding											
Stucco											
Other											

ESTCHECK

Estimate Checklist (Continued)

Task description	Materials			Labor			Sub bids/ Equip. rental	Task total	Notes/ allowances
	Unit	Amt.	Cost	Total	Unit	Amt.	Cost	Total			

Task description	Unit	Amt.	Cost	Total	Unit	Amt.	Cost	Total	Sub bids/ Equip. rental	Task total	Notes/ allowances
Decorating											
Decorator fee											
Model home furnishing											
Other											
Landscaping											
Bark/gravel/rock											
Boxes /borders/ pots											
Flowers											
Fountain											
Grading											
Irrigating											
Shrubs											
Sod/ lawn											
Topsoil											
Trees											
Yard light											
Other											
Misc. & other structures											
Breezeway											
Decking											
Fence - block @ ft.											
Fence - chainlink @ ft.											
Fence - other											
Fence - prefab concrete											
Fence - wood @ ft.											
Final cleanup											
Gazebo											
Glass breakage allow.											
Greenhouse											
Interest incurred											
Lattice											
Ornamental iron											
Railing											

Estimate Checklist (Continued)

Task description	Materials				Labor				Sub bids/ Equip. rental	Task total	Notes/ allowances
	Unit	Amt.	Cost	Total	Unit	Amt.	Cost	Total			
Shed											
Signs											
Theft											
Vandalism											
Other _____											
Other _____											
Other _____											
Subtotals:											

Contingency _____ %

Overhead _____ %

Profit _____ %

Total estimate _____

Specialty Selection Sheet

Date: _____

Name: _____

Phone number: _____

Project name: _____

Project number: _____

Room	Description	Number	Color	Notes
Roofing				
Siding				
Kitchen	Cabinets			
	Countertop			
	Flooring			
	Appliances			
Bath 1	Fixtures			
	Countertop			
	Flooring			
Bath 2	Fixtures			
	Countertop			
	Flooring			
Laundry				
Carpet				

I agree to the selections made above and understand that if I wish to make any changes, the Builder has the right to deny the request or add a service charge to the cost of the project to cover any costs incurred in making the changes.

Client: _____

Signed: _____   Date: _____

SPECSEL

Project Management Contract

Date: _____
Owner: _____
Project name: _____
Project number: _____

This agreement is made this _____ day of _____, 19____ by and between _____ _____, hereinafter called the Owners, whose address is _____ _____, and _____, hereinafter called the Project Manager, Contractor's License No. _____, whose address is _____ _____.

The parties above hereby agree as follows:

The Project Manager agrees to manage and see to completion in an excellent and substantial manner, the construction of a single family residence (herein called the "Project") for _____ _____ upon the real property located at _____ .

The Project Manager's Roles and Responsibilities are to be as follows:

Pre-Construction Phase

1. Assist in finding a lender;

2. Obtain all necessary permits and handle all interface with the building department (general, grading, electrical, mechanical, plumbing, etc.);

3. Provide a complete materials takeoff (bill of materials) for lumber and framing hardware;

4. Coordinate the bidding process with at least three bids per each major craft;

5. Act as the clearing house for information to bidding subs;

6. Check up on the references of each subcontractor;

7. Assemble presentation to the lender;

8. Provide a detailed cost breakdown sheet for the lender;

9. Develop a critical path time line with completion of the living space targeted for _____ ;

10. Act as the contractor of record with the lender;

11. Recommend to the Owners the most qualified and reasonably priced subcontractor for each phase of construction based on the bids submitted;

12. Finalize the contract documents with each sub, making sure everything is "spelled out";

13. Develop a list of contract addenda covering areas of payment disbursement, safety, workers' compensation, cleanup, and craftsmanship standards.

Project Management Contract (Continued)

Construction Phase

Goal: Make sure something progresses on the job site every working day in order to get the _____ _____ into their new home by _____.

The Project Manager will see to it that the Project is constructed and completed in strict conformance with the plans and specifications and that all laws, ordinances, rules, and regulations of the applicable governmental authorities are adhered to. Further, he agrees to:

1. Be the Owners' "eyes and ears in the field";
2. Meet with Owners on a regular basis to review progress;
3. Coordinate all utilities hookups;
4. Coordinate with the geologist and/or soils engineer;
5. Coordinate with the grader;
6. Supervise day-to-day construction, making sure the Project is built as intended by the Owners and the Designer as per plans;
7. Make sure neighbors' property is respected;
8. Ensure subs maintain the highest degree of craftsmanship;
9. Schedule subs and suppliers;
10. Review workers' comp. of each sub — Have a Certificate of Workers' Comp. sent to Project Manager by their carrier;
11. Coordinate incidental day labor;
12. Keep track of change orders;
13. Verify lumber and materials deliveries to make sure we get what we ordered;
14. Authorize payment of subs based upon ongoing review of their work;
15. Obtain lien releases;
16. Provide on-site problem-solving with subs;
17. Coordinate with interior designer's installation;
18. Call for and handle inspections and interface with Building Department;
19. Make sure the job site is cleaned on a regular basis (Owners to provide manpower or funds for labor);
20. Insist that all workers maintain the highest safety standards;
21. Pass along to the Owners any discounts the Project Manager is entitled to as a General Contractor without charging any markups.

Project Management Contract (Continued)

The Owner's Roles and Responsibilities are to be as follows:

1. Finalize negotiations with a lender and secure the construction loan;

2. Apply for and pay for the General Building Permit. All other permits are to be paid for and taken out by each respective subcontractor;

3. All construction expenses are to be borne by the Owners. The Project Manager is not to be held liable for any unpaid bills;

4. Sign all contracts with each subcontractor upon the recommendations of the Project Manager;

5. Clear through Project Manager any work or changes on the Project;

6. Maintain an account for incidental purchases (petty cash) — Project Manager will keep track of these expenses;

7. Provide a job site phone upon commencement of rough grading (cost included in the construction loan);

8. Provide a temporary construction field office available to the Project Manager upon commencement of rough grading (cost included in the construction loan);

9. Accept the liability for theft or destruction of building materials;

10. Provide ALL the appropriate insurances (public liability, course of construction, fire, theft, etc.);

11. Provide a chain-link fence around the Project area;

12. File Notice of Completion within 5 days of substantial completion of the Project.

Fee Schedule

In consideration for the above services rendered by the Project Manager, the Owners agree to pay the Project Manager a Base Fee of 10% of the cost of construction, based on the amount of the construction loan, plus, as an incentive to bring the Project in under budget, a Final Draw of 50% of any unused loan funds that exist when the Notice of Completion is filed by the Owners and all construction budgeted for in the construction loan is completed.

The Base Fee is to be computed based on the total hard construction costs approved in the construction loan, including permit costs, appliances and fixtures. Any soft costs included in the construction loan, such as design, engineering, interior design as well as furniture and drapery costs, are not to be considered in determining the Base Fee.

The Base Fee is to be broken down into installments as follows:

An initial retainer of $_____ to be paid by _____, to cover the first month of pre-construction services.

A monthly installment of 1/12th of the base fee to be paid on the 1st of each following month up to _____
_____.

Project Management Contract (Continued)

The final draw is to be paid within 15 days of filing the Notice of Completion, based on the following:

* As a mutually-beneficial incentive to finish the construction in a timely manner and to meet the completion goal stated above, the Project Manager shall receive the full 50% final draw as stipulated above if construction of the living space is completed to warrant a Certificate of Occupancy from the City by _____ enabling the Owners to move in by _____.

* If Certificate of Occupancy is not granted by _____, 10% shall be deducted from the Project Manager's final draw for each 15 calendar day increment beyond _____.

* If progress is delayed for any reason beyond the Project Manager's control, the time frame for computing the final draw shall be shifted beyond _____ by the amount of the delay. Examples of such delays include: unseasonably rainy weather, earthquakes, fire, work stoppages and unreasonable delays in city approvals (defined as requests that may delay rough grading beyond _____.

* Cost overruns, in any particular construction budget category or allowance, made by the Owners during construction that are a result of changes or upgrades not called out in the plans and specs, are not to be considered as part of any remaining construction funds for purposes of determining the final draw.

We hereby execute this Agreement in _____ on _____

Owner (s):

_____

_____

_____

Project Manager:

_____

Layout Sheet

Layout date:	Sale date:
Estimate by:	Plans: Y / N
Customer:	
Phone:	Job phone:
Owner at layout: Mr / Mrs / Ms	
Job Superintendent:	

Job address:

Trade	Subcontractor	Date called	Meeting time	OK / By
Permits				
Demolition				
Excavation				
Concrete				
Framing				
Roofing				
Siding				
Windows				
Doors				
Gutters / downspouts				
Plumbing				
Electrical				
H.V.A.C.				
Insulation				
Drywall				
Cabinets				
Surfacing				
Floor cover				
Fencing				
Light fixtures				
Painting				
Debris removal				
Lumber				
Finish materials				

Job Checklist

Review contract _____
Review plan _____
Selections completed _____
Salvage items _____
Pets _____

Schedule made _____
Start date _____
Permit _____
Job sign _____

Directions to job site:

_____
_____
_____
_____
_____

LAYOUTSH

New Job Checklist

Job sale date:	Job #:
Salesperson:	Job superintendent:
Contract amount:	Allowance amounts:

Customer name / address:

Manager review:
- Contract & job # _____
- Down payment _____
- Allowances _____
- Selections _____
- Estimate _____
- Layout sheet _____
- Specialty items _____
- Thank you letter _____

_____
(Manager OK)

Bookkeeping:
- Check all math _____
- Check pay schedule _____
- Job set up _____

_____
(Bookkeeping OK)

Job Superintendent:
- Read contract _____
- Check estimate _____
- Check allowances _____
- Check selections _____
- Check special orders _____
- Check layout sheet _____
- Confirm subs at layout _____
- Schedule permit appointment _____
- Schedule plans _____
- Confirm start date _____

_____
(Superintendent OK)

Salesperson:
- Thank you note _____
- Confirm layout date w/owner _____
- Check all forms complete _____

_____
(Sales OK)

Projected job start date: _____ / _____ / _____

Projected job completion date: _____ / _____ / _____

Job Schedule

Customer name / address:		Job start date:	Job completion date:
		Job type:	

Month

Date	
General Conditions	
Demolition / tear out	
Excavation	
Concrete	
Masonry	
Framing	
Roofing	
Siding	
Windows	
Doors	
Sheet metal	
Plumbing	
Electrical	
H.V.A.C.	
Insulation / weatherstripping	
Drywall / plaster	
Ceiling tile	

SCHEDUL

Job Schedule

Customer name / address:

Job start date: | **Job completion date:**

Job type:

Month

Date
Cabinets
Surfacing
Tile
Floor covering
Kitchen & bath accessories
Awning & patio
Finish carpentry
Hardware & metalwork
Paneling & fence
Light fixtures
Paint & decor
Debris removal
Miscellaneous
Inspections
Progress payment
Punch list

Change Order No. _____

Date:	Owner:
Job number:	Job phone:
Original contract number:	Original contract date:

Job name / location:

Change (add or delete) the following work to the original contract:

Change the original contract amount by: _____
Previous contract amount: _____
Revised contract amount: _____

We agree to furnish labor & materials complete in accordance with the above specifications at the price stated above.	Above additional work to be performed under the same conditions as specified in the original contract unless otherwise stipulated.
General Contractor Date	Owner Date

Note: This change order becomes part of the original contract.

CHANGE

Subcontractor Change Order

Date:	
Owner:	
Contractor:	
Project name:	
Change order number:	
Original contract date:	

You are directed to make the following changes in this contract:

_____
_____
_____
_____
_____
_____
_____
_____
_____

The original contract sum was: $ _____

Net amount of previous change orders: _____

Total original contract amount plus or minus net change orders: _____

Total amount of this change order: _____

The new contract amount including this change order will be: _____

The contract time will be changed by: _____ ( ) Days

The date of completion as of the date of this change order is: _____

Contractor:

_____
Company name

_____
Address

_____
City, State, ZIP

_____
By

_____
Date

Subcontractor:

_____
Company name

_____
Address

_____
City, State, ZIP

_____
By

_____
Date

SUBCONTC

Delay Notice

Date: _____

Owner: _____

Contractor: _____

Project number: _____

Project name: _____

o: _____

ttention: _____

Notice number: _____

Contract date: _____

_____

_____

is apparent that you are causing delay to the project. We request that you do whatever is necessary to get
ack on schedule, including but not limited to:

_____	Overtime work
_____	Adding additional crew
_____	Better organization of efforts
_____	Using more experienced personnel
_____	Supplying needed tools and materials

ote:

_____

_____

_____

_____

_____

_____

_____

_____

By: _____   Copy to: _____

Potential Backcharge Notice

Date: _____
To: _____
Notice number: _____
Contract number: _____
Project number: _____
Project name: _____

It is apparent that we will incur additional expense due to the following:

- ☐ Correcting defective work
- ☐ Finishing uncompleted work
- ☐ Damage to finished surfaces
- ☐ Debris removal
- ☐ Other: _____
- ☐ Other: _____

- ☐ Please complete the work noted in accordance with the Contract so backcharges can be avoided.
- ☐ Estimated backcharge cost $_____
- ☐ Please refer to Article _____ of the Contract

Note:

_____
_____
_____
_____
_____
_____
_____
_____
_____
_____
_____
_____
_____
_____
_____
_____

Final Completion List

Customer:

Address of Job:

#	Item	Work to be done	Date complete	Owner approved
	General conditions			
	Demolition / tear out			
	Excavation			
	Concrete			
	Masonry			
	Framing			
	Roofing			
	Siding			
	Windows			
	Doors			
	Sheet metal			
	Plumbing			
	Electrical			
	H.V.A.C.			
	Insulation / weatherstripping			
	Drywall / plaster			
	Ceiling tile			
	Cabinets			
	Surfacing			
	Tile			
	Floor covering			
	Kitchen & bath accessories			
	Awning & patio			
	Finish carpentry			
	Hardware & metalwork			
	Paneling & fence			
	Light fixtures			
	Paint & decor			
	Debris removal			
	Miscellaneous			
	Inspections			

Final completion date: ____ / ____ / ____

_____

(Owner)

JOBCOMPL

Project Closeout Checklist

Date: _____
Owner: _____
Contractor: _____
Project number: _____
Project name: _____

The following is a checklist of items that should occur when finalizing a project:		Construction manager	Site superintendent	Foreman
1.	Coordinate final utility and service connections.			
2.	Complete acceptance checklists.			
3.	Arrange for final agency inspection.			
4.	File Notice of Completion.			
5.	Thoroughly clean the project.			
6.	Conduct an acceptance walk-through with the contractors and with the client.			
7.	Process final payment from client or lender.			
8.	Assist in systems start-up.			
9.	Obtain warranties and lien releases from subs.			
10.	Create owner's manual with warranties and project literature.			
11.	Procure as-built drawings.			
12.	Obtain maintenance and operating instructions.			
13.	Initiate preventive maintenance program as appropriate.			
14.	Assist in teaching operating and maintenance staff when needed.			
15.	Obtain and store spare parts and materials as required.			
16.	Dispose of excess materials.			
17.	Remove temporary facilities, tools and equipment.			
18.	Recap the actual schedule and job costs for the data bank.			
19.	Remove data from computer memory after all cost entries have ceased.			
20.	Transfer and obtain sign-off for keys.			
21.	Obtain letter of recommendation from client.			
22.	Follow up with client survey 30 days after close of job.			

PROJCLOS

Project Closeout Checklist (Continued)

	Construction manager	Site superintendent	Foreman
3.			
4.			
5.			
6.			
7.			
8.			
9.			
10.			
11.			

Monthly Sales Report

Month: _____

Salesperson: _____

Date	Customer	Callbacks	Quote	Sale Yes / No / Pending	Notes
____	_____	_____	____	Y N P	_____
____	_____	_____	____	Y N P	_____
____	_____	_____	____	Y N P	_____
____	_____	_____	____	Y N P	_____
____	_____	_____	____	Y N P	_____
____	_____	_____	____	Y N P	_____
____	_____	_____	____	Y N P	_____
____	_____	_____	____	Y N P	_____
____	_____	_____	____	Y N P	_____
____	_____	_____	____	Y N P	_____
____	_____	_____	____	Y N P	_____
____	_____	_____	____	Y N P	_____
____	_____	_____	____	Y N P	_____
____	_____	_____	____	Y N P	_____
____	_____	_____	____	Y N P	_____
____	_____	_____	____	Y N P	_____
____	_____	_____	____	Y N P	_____
____	_____	_____	____	Y N P	_____
____	_____	_____	____	Y N P	_____
____	_____	_____	____	Y N P	_____
____	_____	_____	____	Y N P	_____
____	_____	_____	____	Y N P	_____
____	_____	_____	____	Y N P	_____
____	_____	_____	____	Y N P	_____
____	_____	_____	____	Y N P	_____
____	_____	_____	____	Y N P	_____
____	_____	_____	____	Y N P	_____
____	_____	_____	____	Y N P	_____
	Total:		____	__ __ __	

New leads: _____

Leads contacted: _____

Sales: _____

Pending: _____

Ratio of leads to sales: _____

Total sales for month: _____

Commission earned: _____

Quarterly Overhead Review
General Contractor – New Home Construction
(Computations based on annual volume built and collected)

Overhead Item	Low / High Percents	Last Quarter	Percent	This Quarter	Percent
1. Advertising	1.00 to 2.00				
2. Sales	2.00 to 5.00				
3. Office Expenses					
Staff	1.50 to 3.50				
Rent	0.35 to 1.20				
Office equipment	0.10 to 0.50				
Telephone	0.10 to 0.75				
Computer	0.10 to 0.35				
Office supplies	0.05 to 0.20				
4. Job Expenses					
Vehicles	0.75 to 2.50				
Job supervision	1.50 to 2.50				
Tools & equipment	0.15 to 0.50				
Service & callbacks	0.20 to 0.40				
Mobile telephone	0.05 to 0.30				
Pagers	0.05 to 0.15				
5. General Expenses					
Owner's salary	4.00 to 6.00				
General insurance	0.25 to 1.00				
O.C.R.A.	1.00 to 4.00				
Interest	1.75 to 3.00				
Taxes	0.00 to 3.00				
Bad debts	0.00 to 0.15				
Licenses & fees	0.10 to 0.25				
Accounting fees	0.10 to 0.25				
Legal fees	0.15 to 0.30				
Education & training	0.10 to 0.20				
Entertainment	0.05 to 0.10				
Association fees	0.10 to 0.10				
Totals	15.45% to 38.20%				

Quarterly Overhead Review
General Contractor – Remodeling
(Computations based on annual volume built and collected)

Overhead Item	Low / High Percents	Last Quarter	Percent	This Quarter	Percent
1. Advertising	1.50 to 5.00				
2. Sales	5.00 to 8.00				
3. Office Expenses					
Staff	3.00 to 7.00				
Rent	0.35 to 1.20				
Office equipment	0.10 to 0.50				
Telephone	0.20 to 0.75				
Computer	0.10 to 0.35				
Office supplies	0.05 to 0.20				
4. Job Expenses					
Vehicles	0.75 to 3.00				
Job supervision	4.00 to 6.00				
Tools & equipment	0.20 to 0.75				
Service & callbacks	0.10 to 0.50				
Mobile telephone	0.05 to 0.30				
Pagers	0.05 to 0.15				
5. General Expenses					
Owner's salary	6.00 to 8.00				
General insurance	0.25 to 1.50				
O.C.R.A.	1.00 to 4.00				
Interest	0.50 to 0.75				
Taxes	0.00 to 3.00				
Bad debts	0.00 to 0.30				
Licenses & fees	0.10 to 0.25				
Accounting fees	0.15 to 0.30				
Legal fees	0.15 to 0.50				
Education & training	0.15 to 0.30				
Entertainment	0.10 to 0.20				
Association fees	0.10 to 0.20				
Totals	**23.50% to 53.00%**				

Quarterly Overhead Review
Specialty Contractor
(Computations based on annual volume built and collected)

Overhead Item	Low / High Percents	Last Quarter	Percent	This Quarter	Percent
1. Advertising	1.00 to 2.00				
2. Sales	1.00 to 2.00				
3. Office Expenses					
Staff	3.00 to 7.00				
Rent	0.35 to 1.20				
Office equipment	0.10 to 0.50				
Telephone	0.25 to 0.75				
Computer	0.10 to 0.35				
Office supplies	0.05 to 0.20				
4. Job Expenses					
Vehicles	0.75 to 3.00				
Job supervision	3.00 to 5.00				
Tools & equipment	0.20 to 0.75				
Service & callbacks	0.10 to 0.50				
Mobile telephone	0.05 to 0.30				
Pagers	0.05 to 0.15				
5. General Expenses					
Owner's salary	6.00 to 8.00				
General insurance	0.25 to 1.50				
O.C.R.A.	1.00 to 4.00				
Interest	0.50 to 0.75				
Taxes	0.00 to 3.00				
Bad debts	0.00 to 0.30				
Licenses & fees	0.10 to 0.25				
Accounting fees	0.15 to 0.30				
Legal fees	0.15 to 0.40				
Education & training	0.15 to 0.30				
Entertainment	0.10 to 0.20				
Association fees	0.10 to 0.20				
Totals	18.50% to 42.90%				

Overhead Expense Chart

(For every $100,000 in annual volume sold, built and collected)

Overhead Item	Low percent	High percent	Low expense	High expense
1. Advertising				
2. Sales				
3. Office Expenses				
Staff				
Rent				
Office equipment				
Telephone				
Computer				
Office supplies				
4. Job Expenses				
Vehicles				
Job supervision				
Tools & equipment				
Service & callbacks				
Mobile telephone				
Pagers				
5. General Expenses				
Owner's salary				
General insurance				
O.C.R.A.				
Interest				
Taxes				
Bad debts				
Licenses & fees				
Accounting fees				
Legal fees				
Education & training				
Entertainment				
Association fees				
Totals				

Profit and Loss Statement

	From _____	to _____	%	Year to date _____	%
Revenue					
Residential remodel income	_____			_____	
Commercial rehabilitation	_____			_____	
Residential development	_____			_____	
Other income	_____			_____	
Total revenue		_____		_____	
Job costs					
Direct labor	_____			_____	
Direct labor burden	_____			_____	
Materials	_____			_____	
Subcontractors	_____			_____	
Equipment rentals	_____			_____	
Dump fees	_____			_____	
Permits	_____			_____	
Total job costs		_____		_____	
Overhead expenses					
Advertising	_____			_____	
Business meals	_____			_____	
Entertainment	_____			_____	
Dues & publications	_____			_____	
Officers' salaries	_____			_____	
Administrative wages	_____			_____	
Overhead labor burden	_____			_____	
Office & shop rent	_____			_____	
Telephone & mobile phone	_____			_____	
Utilities	_____			_____	
Business license & fees	_____			_____	
Office supplies	_____			_____	
Legal & accounting	_____			_____	
Bad debts	_____			_____	
Bank charges	_____			_____	
Gas, oil & repairs	_____			_____	
Vehicle lic., reg., & ins.	_____			_____	
Small tools	_____			_____	
Depreciation	_____			_____	
Total overhead expenses		_____		_____	
Net income or (loss):		_____		_____	

PROF&LOS

Break-Even Point Worksheet

Total projected sales volume for year $ _____

Total projected overhead for year $ _____

Total projected job costs for year $ _____

Break even = $Total overhead ÷ $$\frac{(\$\text{Total sales volume} - \$\text{Total job costs})}{\$\text{Total sales volume}}$$

Break even = $ _____ ÷ $$\frac{(\$\_\_\_\_\_ - \$\_\_\_\_\_)}{\$\_\_\_\_\_}$$

Break even = $ _____ ÷ $$\frac{\$\_\_\_\_\_}{\$\_\_\_\_\_}$$

Break even = $ _____ ÷ $ _____

Break even = $ _____

The break-even point for the year for your company is $ _____.

To find the break-even point for the average month for your company, divide your break-even point for the year by 12:

$ _____ ÷ 12 = $ _____

To compute the approximate date that you will reach your break-even point during the year, divide your projected sales volume for the year by 12 to get your projected sales volume for each month. Then divide your break-even point for the year by your projected sales volume for the month:

$ _____ ÷ 12 = $ _____

$ _____ ÷ $ _____ = $ _____ or _____

                                                                                                       (date)

Appendix B
Educational and Construction Resources

Here are the names and addresses of businesses and associations that provide books, video and audio cassette programs, computer programs and other educational or business materials that you can use to improve your business expertise. I've had dealings with each of them, found them to be very service-oriented and easy to work with, and so feel comfortable recommending them.

Associations

National Association of Home Builders
1201 15th Street, N.W.
Washington, DC 20005-2800

(202) 822-0200
(800) 368-5242
Fax: (202) 822-0559
Internet: www.nahb.com

National Association of the Remodeling Industry
780 Lee Street, Suite 200
Des Plaines, IL 60016

(800) 611-6274
Fax: (847) 298-9225
Internet: www.remodeltoday.com
E-mail: info@nari.org

Recommended Books on Self-Improvement & Sales

The E-Myth*
Michael Gerber (published by Harper Collins Publishers)

The Goal*
Eliyahu Goldratt (published by The North River Press)

How to Master the Art of Selling*
Tom Hopkins (published by Warner Books)

Selling for Dummies*
Tom Hopkins (published by IDG Books)

The 7 Habits of Highly Effective People*
Stephen R. Covey (published by Simon & Shuster)

*May be purchased through most local bookstores

Audio & Video Cassette Programs for Self-Improvement & Sales

CareerTrack
3085 Center Green Drive
Boulder, CO 80301-5408

(303) 447-2323
(800) 334-1018
Fax: (800) 832-9489
Internet: www.careertrack.com

Nightingale-Conant
7300 N. Lehigh Avenue
Niles, IL 60714

(847) 647-0306
(800) 525-9000
Fax: (847) 647-7145
Internet: www.nightingale.com

The Telephone Doctor
30 Hollenberg Court
St. Louis, MO 63044

(314) 291-1012
(800) 882-9911
Fax: (314) 291-3710
Internet: www.telephonedoctor.com

Tom Hopkins International
P.O. Box 1969
Scottsdale, AZ 85252

(602) 949-0786
(800) 528-0446
Fax: (602) 949-1590
Internet: www.tomhopkins.com
E-mail: th@primenet.com

Seminars

CareerTrack
3085 Center Green Drive
Boulder, CO 80301-5408

(303) 447-2323
(800) 334-1018
Fax: (800) 832-9489
Internet: www.careertrack.com

Construction Programs & Results
1001 49th Street
Washougal, WA 98671

(360) 335-1100
(888) 944-0044
Fax: (360) 835-1148
Internet: www.markupandprofit.com
E-mail: michael@markupandprofit.com

The Telephone Doctor
30 Hollenberg Court
St. Louis, MO 63044

(314) 291-1012
(800) 882-9911
Fax: (314) 291-3710
Internet: www.telephonedoctor.com

Tom Hopkins International
P.O. Box 1969
Scottsdale, AZ 85252

(602) 949-0786
(800) 528-0446
Fax: (602) 949-1590
Internet: www.tomhopkins.com
E-mail: th@primenet.com

Appendix B: Educational and Construction Resources

Construction Books

Building Tech Bookstore, Inc.
8020 S.W. Cirrus Drive
Beaverton, OR 97008-5986

(503) 641-8020
(800) 275-2665
Fax: (503) 641-0770
E-mail: mail@buildingtechbooks.com

Craftsman Book Company (see order form in back of this book)
6058 Corte del Cedro
P.O. Box 6500
Carlsbad, CA 92018

(760) 438-7828
(800) 829-8123
Fax: (760) 438-0398
Internet: www.craftsman-book.com

Home Builder Bookstore
National Association of Home Builders
1201 15th Street, N.W.
Washington, DC 20005-2800

(202) 822-0384
(800) 223-2665
Fax: (202) 822-0512
Internet: www.builderbooks.com

Construction Programs & Results
1001 49th Street
Washougal, WA 98671

(360) 335-1100
(888) 944-0044
Fax: (360) 835-1148
Internet: www.markupandprofit.com
E-mail: michael@markupandprofit.com

Appendix B: Educational and Construction Resources

Computer Programs

Craftsman Book Company (see order form in back of this book)
6058 Corte del Cedro
P.O. Box 6500
Carlsbad, CA 92018

(760) 438-7828
(800) 829-8123
Fax: (760) 438-0398
Internet: www.craftsman-book.com

Construction Programs & Results
1001 49th Street
Washougal, WA 98671-9761

(360) 335-1100
(888) 944-0044
Fax: (360) 835-1148
Internet: www.markupandprofit.com
E-mail: michael@markupandprofit.com

Daily Scheduling

DAY-TIMERS, Inc.
One Day-Timer Plaza
Allentown, PA 18195-1551

(800) 225-5005
Fax: (800) 452-7398
Internet: www.daytimer.com

Sales Test

Lousig-Nont & Associates
Industrial Psychologists
3740 S. Royal Crest Street
Las Vegas, NV 89119-7010

(702) 732-8000
(800) 477-3211
Fax: (702) 732-1572
Internet: www.helen-robinson@usa.net

Index

A

Accounting software .... .194
Accounting, understanding 8
Accounts, chart of ...... .189, 190-192
Adjusted job costs .... .204
Adjusting to new ideas .... .12
Advertising
  business cards 90-93
  customer referral .... .96
  direct mail 95-96
  door hangers .... .94
  importance to sales .... .89
  job signs 93-94
  lead time for sales .... .173
  Yellow Page .... .94
  word-of-mouth .... .63
Advice, legal 125-127
Agenda, meeting .... .216
Allowance limits .... .29
Allowances, low balling .... .119
Anti-education syndrome .... .22
Appointments, keeping .... .99
Architects, dealing with 112-114
Arguments with customer .... .159
Asset to liability ratio .... .188
Assimilating an unbudgeted
  expense 72-77
Associations, participation in ..... .222
  addresses .... .305
Attitudes
  adjusting yours .... .12
  employee .... .215
  successful .... .10
  toward customers .... .98
Attorney
  selecting 125-126
  writing contracts for 136-137
Audio tapes, as educational
  tools .... .22, 306
Author biography .... .317
Automating with computers ..189, 192

B

Backcharge Notice, See page 265
Bad checks .... .133
Bank owned property 202-203
Bankers
  choosing yours 200-201
  for customers 199-200
  referrals to/from .... .201
  working for 202-203
Basic steps of sale 100-112
Basics for survival 7-9
Bean counters .... .197
Bidding .... .24
  contests .... .226
Bids
  low .... .25, 225-226
  on architect's plans .... .112
  written 115-116
Blank forms 265-304
  installing from CD-ROM ..265-267
Bonus clause, contract .... .135

Bookkeeping
  separating work types .... .61
  staff 198-199
  systems .... .199
Break-even chart
  new home construction .... .183
  remodeling .... .182
Break-even date, calculating ..... .182
Break-even point
  defined .... .180
  formula .... .181
  problem to solve 232-235
Break-Even Point Worksheet,
  See page 265
Brokers, choosing 201-202
Budget
  architectural changes in .... .114
  establishing customers' 106-112
  for O.C.R.A. .... .176
  setting your .... .169
  working within 16-17
Budget adjustments
  decreasing profits .... .75
  importance of .... .71
  increasing markup .... .74
  increasing sales .... .72
  more than one per year .... .76
  spread over more than one year ...74
Building departments, dealing
  with 197-198
Business cards 90-93, 214
Business cycles 62-63
Business expenses .... .27
Business failure 20-26
  falling behind the industry .... .21
  losing contact with business ..... .22
  major causes .... .24
  not recognizing problems .... .20
  owner responsibility .... .20
Business knowledge, importance of .11
Business maturity .... .225
Business organizations .... .222
Business planning .... .24
Business problems, evaluating 251-261
Business trends, keeping up with ..211
Business vs. personal
  expenses 251-257
Buying decisions, customer ...... .103
Buying equipment .... .185

C

CAD programs .... .194
Calculating markup 53-54
Calculating work volume .... .40
Calculations
  decreasing profits for unbudgeted
    expense .... .75
  increasing markup for unbudgeted
    expense .... .74
  increasing sales for unbudgeted
    expense .... .72
  second budget adjustment ...... .76
  sliding scale markup .... .69
Callbacks, to set budget 110

Capital, projections for .... .24
Capital reserve account ..170, 175-178
Cash flow
  during down times .... .63
  generating .... .172
  projections for .... .24, 170-174
Cash reserve, importance of ..... .178
Casualty work 203-206
CD-ROM, using forms on 265-267
Cell phone, importance of .... .207
Change Order, See page 265
Change order .... .114, 129
  credit due .... .153
  importance of consistency ...... .154
  markup on 155-158
  payment for .... .152
  purchasing forms .... .149
  sales commissions on .... .155
  sample .... .151
  using .... .150
Change Order, Subcontractor,
  See page 265
Chart of accounts ...... .189, 190-192
Checklist for making money ...... .14
Checklists .... .164, 165
  Estimate, See page 265
  New Job, See page 265
  Project Closeout, See page 265
Checks, bounced .... .133
Classes, continuing education 211
Closeout Checklist, See page 265
Commercial work
  gross profit ratio .... .187
  pricing .... .66
  volume built per employee ...66-67
Commissions, tracking ...... .160, 162
Communication
  office 198-199
  written .... .208
Community involvement .... .214
Commuting, charging for .... .184
Company funds, misuse of .... .23
Company meetings 215-216
Company policies .... .13
Company vs. personal assets ..... .23
Compensation, mileage .... .184
Competency, level required 7
Competitive bidding .... .24, 226
Competitive bids, specialty trades ..65
Completion List, See page 265
Computer
  automating company ...... .189, 193
  drawing programs .... .194
  software 193-196, 308-309
  what to buy .... .192
Computer generated forms ...... .164
  Profit & Markup CD-ROM .265-267
Computer programs, employee
  friendly .... .195
Consultants, hiring 223-224
Consumer protection laws .... .140
Consumer research study, buying
  trends .... .32
Contacts, telephone .... .97
Contacts to leads ratio .... .92

Index

Contract, Project Management,
   See page 265
Contractor as teacher 7
Contractor failure rate 5, 19
Contractor responsibility, contracts 136
Contractors, anti-education
   syndrome 22
Contracts 121
   adjusting for price changes 114
   changes to 129
   contractor solely responsible
      clause 136
   for attorneys 136-137
   handyman 143-145
   importance of detail 122
   job supervision 86
   language, importance of 128
   legal advice on 127
   one page 121
   owner/customer involvement
      on job clause 128, 131
   packaging 123
   payment schedules 132
   penalty clause 135
   penalty for bounced checks . 133-134
   pets on job site 130
   retainage clause 134
   sample 138-139
   software programs for
      . . . . 143, 146-147, 194
   subcontractor 140-142
   supervision work 131-133
   using form documents 127
   writing 126-147
Conventions, construction 211
Cost-of-sales ratio 188
Cost plus estimating 83-85
Costs, debating with customer . . . 159
Costs, defined 26
CPA, advice from 199
Credit due, change orders 153
Crews, keeping work for 63
Current ratio 188
Customer etiquette 98-100
Customer response surveys 211
Customers
   charging for travel 184
   establishing budgets 106-112
   evaluating needs of 101
   explaining change orders to
      . . . . 129, 150-153
   financing for 199-200
   getting a commitment from 108
   good communication with 207
   improving relations with 211
   involvement on job 128, 131
   keeping appointments with 99
   listening to 97
   meeting with 98
   negotiating price 81
   objections to change order costs
      158-159
   pets on job site 130
   problem 226

qualify for buying 103
right of cancellation 123-124
setting job start date 103
third-party involvement 105

D

Database, building your own 195
Dawson, Roger, *Power Negotiating
   for Sales People* 89
Day Timer 227, 309
Death of partner, problem to
   solve 251-260
Deductibles, insurance work 205
Definitions
   industry terms 26-27
   margin 179
Delay Notice, *See page 265*
Depreciated amounts, insurance
   work 205
Direct mail advertising 95-96
Discounting work 242-246
Disk, forms on 265-267
Documents
   contract 123
   form 127
Dollar volume
   construction per household 39
   per employee 43-47
   projecting 171
Door hangers 94
Down payments 119-120
   consumer protection laws against 140
Drawing programs, computer 194
Draws, contract 139-140
Dress standards, sales staff 98
Drive-time learning 22
Dun and Bradstreet, report on
   business failures 19

E

Earnest money 120
Economic levels, customer . . . 109-110
Education
   classes, seminars, conventions . . . 211
   commitment to 8, 11
   importance of good spelling . . . 208
   keeping up with the industry 21
Educational tools 22
Employee
   business leads 91
   commuting to job site 184
   company meetings 215-216
   down time business cycles . . . 62-63
   dress standards 98
   job training 213
   markup training 217-218
   positive encouragement for 210
   productivity 214-215
   ratio to dollar volume 42-45
   resistance to change 45
   using change orders 129
   using computer software 195

Equipment, rent or buy 185
Error factors, estimating 28
Escrow accounts 140
Estimate Checklist, *See page 265*
Estimate Form, *See page 265*
Estimates, low ball 118
Estimating
   computer software 194
   insurance work 205-206
   job costs 27-28
Estimating methods 28-32
   cost plus 83-85
   guessing 29
   stick 29, 31
   time and materials 83-85
   time involved 31
   unit cost 29-30
Etiquette, customer 98-100
Evaluating expenses, problem
   to solve 247-261
Evaluating customer needs 101
Evaluating proposals, problem
   to solve 242-246
Expenses
   adjusting for, problem to
      solve 236-241
   budgeting for new 72-77
   fixed and variable 38
   labor 170
   overhead 32-37
   projecting 171
   up-front job costs 120
Expert, industry 219-220
   becoming 221

Failure, business
   among contractors 5, 19
   keeping up with industry 21
   losing contact with business 22
   major causes of 24
   owner responsibility 20
   symptoms of 20-26

F

Family member's salary 48
Fast Track Proposal Writer 143
Fax, contracts 143, 146
Fax cover sheet 146
Fax machine, importance of 196
Field staff, using change orders . . . 129
Final Completion List, *See page 265*
Financing, customer 199
Five basics for success 7-9
Fixed overhead 38
Focus, importance in business
   . . . . 15, 227-228
Form documents
   computer generated 164
   installing from CD-ROM . . . 265-267
   purchasing preprinted 149
   suggested for use 160-167
   using 127
Forms, blank, *See page 265*

Index

Forms, sample
  Final Completion List 148
  Job Schedule 166-167
  Layout Sheet 163
  Monthly Sales Report 162
  New Job Checklist 165
  Overhead Expense Charts 35-37
  Quarterly Overhead Review,
    New Home Construction 50
  Quarterly Overhead Review,
    Remodeling 49
  Quarterly Overhead Review,
    Specialty Contractor 51
  Quotation Request 161
Formula, using incorrect for
  markup 82
Formulas
  assimilate unbudgeted expense . . .77
  break even 181
  cost-of-sales ratio 188
  current ratio 188
  gross profit ratio 187
  job cost adjuster 204
  job costs .56
  marketing expense ratio 188-189
  markup 18, 56
  overhead percentage 56
  profit 56
  sales price 57
  total volume built 55, 59
  travel time 183-184
  verify markup 55
Funding, long term 24

G

General contractors
  competitive bids 65
  working with specialty trades . .64-65
Gerber, Michael, *The E-Myth* . . .7, 306
Goals
  focusing on 227-228
  production 54
  setting 13-14
  tracking 181
Government agencies, contracts
  with 134
Government work, low bids 225
Grammar and speaking 209-210
Gross margin 179
Gross profit, per manhour 187
Gross profit ratio 186-187
Groups, networking 222-223

H

Handyman contract 143-145
Hangers, door 94
Help, in your business 222-224
Helping others 221-223
High-pressure sales 108
Hiring consultants 223-224
Honesty, importance of 9
Hopkins, Tom, sales trainer
  . . . . .15, 306-307

I

Increases in labor/materials,
  calculating 242-246
Indirect expenses 27
Industry experts
  becoming one 221
  listening to 34, 38
  recognizing 219-220
Industry terminology 26
Installing *Profit & Markup* CD-ROM
  on Macintosh 266
  on *Windows*™ 265
Insurance companies, low bids 225
Insurance work 203
  deductibles on 205
  estimates for 205-206
  markup on 204
Intuit's *QuickBooks* 194

J

Job changes, customer 150-153
Job Checklist, *See page 265*
Job checklists 164-165
Job completion punch list 147-148
Job cost adjuster, formula 204
Job costing software 194
Job costs
  calculating 56
  defined 26
  estimating 28
  projecting, problem to solve .232-234
Job Estimate Form, *See page 265*
Job price
  calculating 236-241
  establishing with markup 54-55
Job pricing, the "new angle" 117
Job-related expenses 27
Job requirements, commuting 184
Job Schedule 166-167
Job Schedule Form, *See page 265*
Job signs 93
Job superintendent
  training on markup 217-218
  writing work orders 153, 155
Job supervision
  how to charge for 85
  markup for 87
  writing a contract for 86
Job tracking 164, 166-167
Job training, employee 213

Jobs, discounting 242-246
Journeyman's level of competency . . .7

L

Labor costs, evaluating increases
 242-246
Labor expense 170
Language, contract 126-127
Lawsuits, dealing with 251-261
Lawyer, selecting 125-126
Layout Sheet, example 163

Layout Sheet, *See page 265*
Lead Form, *See page 265*
Lead slips 160
Lead time
  advertising 173
  projecting for cash flow 172-173
Lead to sale ratio 173-174
Lead tracking, software 194
Leads
  attracting 89
  employee 91
  generating 173
  ratio to contacts 92
  using lead slips 161-162
Lecturers, qualified 220
Lecturing for others 221
Legal advice, contracts 127
Lenders
  bargaining with 178
  dealing with 198
  finding for customers 199-200
  referrals to/from 201
Letters, customer 208-210
Level of competency, jouneyman 7
Listening
  as a skill 97
  evaluating needs of customer 101
Loading forms from CD-ROM .265-267
Loans, customer 201-202
Location, office 41
Losses, raising markup to
  recover 79-81
Low-ball estimates 118
Low bids .25
  from specialty trades 65
  just say no to 225
  promised rewards for 115

M

Macintosh, installation of
  Markup & Profit 266-267
Magazines, as educational tools 22
Making money, checklist for 14
Management Contract, *See page 265*
Manual, method of operation 13
Margin .178
  calculating 179
  defined 179
  gross profit 186-187
  problem to solve 230-231
Margin percentage 179
Marketing expense ratio 188-189
Markup
  adjusting for competition 261
  adjusting to suit others 78-79
  calculating for new home
    construction 59
  calculating for remodeling business 56
  checking results of 187
  defined 179
  establishing sales price 54-55
  failure to understand 24

for job supervision work 87
formula 18
increasing for unbudgeted expense 74
insurance work 203-204
justifying 207-211
lowering 78, 247-250
negotiating 79
on change orders 155-158
problems to solve
. . . . .230-233, 236-238, 251-259
raising to recoup a loss 79-81
relationship to overhead 71
sliding scale 67-69
the "new angle" 117
training employees 217-218
understanding 8-9
using formula 53-54
using wrong formula 82
Markup & Profit CD-ROM
installation 265-267
Material costs, evaluating
increases 242-246
Material list, stick estimating 31
Maturity, business 225
Meetings, suggestions for 216
Method of Operation Manual 13
Mileage formula 183-184
Model home, office in 41
Money
checklist for making 14
discussing with customer 107
earnest 120
Monthly Sales Report, example . . .162
Monthly Sales Report, *See page 265*
Mortgage brokers
choosing 201-202
for customers 199-200
Motivational training, Zig Ziglar . .8, 9

N

National Association of Home
Builders 305
consumer study 32
Negotiating favorable proposals
. 242-246
Negotiating price 81, 226
Net margin 179
Net profit 48, 186
Networking 222-223
New angle, sales 117
New home construction 176
break-even chart 183
combining with remodeling . . .59-61
commodity business 60, 70
gross profit ratio 187
Overhead Expense Chart, example 36
Overhead Expense Chart,
See page 265
problem to solve 236-241
Quarterly Overhead Review,
example 50
Quarterly Overhead Review,
See page 265

sample markup problem 57-59
ten percent myth 61
volume built per employee 58
New Job Checklist, example 165
New Job Checklist, *See page 265*
Non-productive employees . . .214-215
Northwest Construction Software .146
address 311
Notice, Backcharge, *See page 265*
Notice, Delay, *See page 265*
Notice of Right of Cancellation
. 123-124

O

O.C.R.A., setting up
for new home construction 176
for remodeling company 175
Objections
countering on change orders .158-159
sales 111, 116-117, 122
Office
buying building 261-262
calculating expenses for 42
locating 41
Office manager salary, calculating .239
Office workers 198
On time
importance in construction 64
meeting customers 99
One call closes 115
One-page contracts 121
Open mind, importance of 12
Operating Capital Reserve Account
(O.C.R.A.) 170
importance of 178
setting up 175-177
Organizations, participation in 222
Overhead
defined .27
monthly 171-172
relationship to markup 71
start up .39
Overhead and profit
markup .31
ten percent myth 34
Overhead Expense Chart, examples
new home construction 36
remodeling 35
specialty contractor 37
Overhead Expense Charts, *See page 265*
Overhead expenses
budgeting new 71
combining 33
comparison to dollar volume
built 38, 59, 70
determining 32
fixed and variable 38
high and low percentages 34
non-productive employees . . .214-215
projecting 171
projecting, problem to solve .232-233
reducing, problem to solve . .247-250

specialty contractors 63
spread over more than one year . . .74
taxes 34
Overhead percentage, calculating . . .56
Owner, contract restrictions on 131-133
Owner, business
importance of work involvement . .23
responsibility for failure 20
salary 17, 47
value of time 185

P

Pager, importance of 207
Partner, death of, problem to
solve 251-260
Pay
family member 48
owner 47-48
Payment
change orders 152
customer 133
Payment schedules
supervision work 132
contract 137, 139-140
Payroll expense 170
Penalty clauses 133, 135
Percent of margin, problem to solve
. . . . .230-234, 251-259
Personal expenses, separating out
. 251-257
Personal property, owners' 23
Peterson, Eugene 170
Pets 130
Phone
answering 97
etiquette 207-208
Piggyback computer programs 195
Planning
business 24
daily 227
Plans, contractor solely responsible
clause .136
Policies, company 13
Potential Backcharge Notice,
See page 265
Power Negotiating for Sales People,
Roger Dawson 89
Preparation, meeting customer 100
Price
adjusting for changes . .114, 156-158
debating with customer 159
negotiating 81, 226
Price, job, the "new angle" 117
Problems to solve 229-263
A Fourth-Year Business 232-235
All That Glitters Is Not Gold 242-246
Boom & Bang Construction .251-261
Correcting a Disaster 236-241
Evaluating a Quarterly Review
. 247-250
Getting Started 230-231
Production
employee 43-47, 214-215

Index

goals, importance of 15, 54
gross profit per manhour 187
   in commercial work 67
   increasing 45
   schedules 164, 166-167
Production per employee 44-47
   new home construction 58
   problem to solve 259
   remodeling 43
   specialty contracting 43
Profit
   calculating 56
   decreasing for unbudgeted
      expense 75
   defined 27
   evaluating in proposals 242-246
   gross 186-187
   importance in business 20
   net 186
   projecting 171
   projecting, problem to solve .232-233
   ten percent myth 34
Profit and Loss Statement,
   See page 265
Profit margins, defined 179
Profitable sales, lack of 24
Programs, computer .193-194, 308-309
   Markup & Profit 265-267
Project Closeout Checklist,
   See page 265
Project Management Contract,
   See page 265
Projected volume/overhead/profit/costs,
   problem to solve 232-233
Projections
   cash flow 24, 170-174
   dollar volume 171
   large purchases 261
   work volume 40
Proposal writing, software 194
Proposals, evaluating 242-246
Punch list, sample 148
Purchases, change orders for 153

Q

Qualified Remodeler magazine 7
Qualifying customers 103, 106
Quality vs. good job 15
Quarterly Overhead Review 48
   evaluating, problem to solve .247-250
   new home construction example . .50
   remodeling example 49
   specialty contractor example 51
Quarterly Overhead Review,
   See page 265
QuickBooks 194
Quotation request 163
Quotation Request Form, *See page 265*
Quotes, written 115-116

R

R.E.O. work 202-203
Ratio
   asset to liability 188
   contacts to leads 92
   cost of sales 188
   current .188
   employee to dollar volume 42-45
   gross profit 186-187
   lead to sale 173-174
   marketing expense 188-189
Reading, for improvement .8, 221, 306
Referral work, specialty contractors .63
Referrals
   architect 113
   customer 96
   to/from lenders 201
Remodeling
   as service business 60, 70
   break-even chart 182
   combining with new home
      construction 59-61
   employee study 43
   gross profit ratio 187
   Overhead Expense Chart, example 35
   Overhead Expense Chart,
      See page 265
   Quarterly Overhead Review,
      example 49
   Quarterly Overhead Review,
      See page 265
   sample markup problem 55-56
   setting up an O.C.R.A. 175
   survey on costs 109
   volume built per employee 43
Rentals, tools and equipment 185
Repair or Replacement Services
   contract 144-145
Repair work, insurance 204
Repossessed property,
   working on 202-203
Retainage clause 134
Right of Rescission 123-124

S

Salary
   office manager 239
   owner 17, 47
   spouse 48
Sales
   attracting 89
   increasing for unbudgeted
      expense 72, 77
   monthly reports 162
   projections 172
   ratio to leads 173-174
Sales commissions, on change
   orders .155
Sales expenses, specialty
   contractors 63
Sales meetings 216
Sales price
   calculating 57
   calculating using margin . . .179-180
   establishing with markup 54-55
   markup formula 18
   raising 25
Sales process, understanding 8
Sales Report, *See page 265*
Sales staff
   dress standards 98
   training on markup 217-218
Sales training, Tom Hopkins 15
Sales volume
   comparison to overhead . . .38, 59, 70
   projecting, problem to solve 259
Salespeople
   commissions on change orders . .155
   figuring markup on changes .156-158
   tracking sales 160, 162
Sample contract
   Repair or Replacement
      Services 144-145
   Standard Contract/Proposal .138-139
   Subcontractor 141-142
Sample form
   Change Order 151
   Chart of Accounts 190-192
   Final Completion List 148
   Job Schedule 166-167
   Layout Sheet 163
   Monthly Sales Report 162
   New Job Checklist 165
   Quotation Request 161
Sample problem
   break-even point 181-182
   calculating margin 180
   decreasing profits for unbudgeted
      expense 75
   increasing markup for unbudgeted
      expense 74
   increasing sales for unbudgeted
      expense 72
   markup for new home
      construction 57-59
   markup for residential
      remodeling 55-56
   second budget adjustment 76
   setting up an O.C.R.A. for new
      home construction 176
   setting up an O.C.R.A. for
      remodeling 175
   sliding scale markup 68
   using the wrong formula 82
Sauer, Dave, *Qualified Remodeler* . .7
Schedules
   daily 227
   Job, *See page 265*
   keeping up with 64
   work 164, 166-167
Scheduling start date 103
Selection Sheet, *See page 265*
Self-improvement, steps for . .221-222
Selling
   basic techniques of 100-112
   commitment to price 81
   evaluating needs of customer 101

forms needed 161-163
getting a commitment to buy 108
getting down payments 119-120
high pressure 108
knowing when to walk away 105
low-balling estimates 118-119
meeting customer objections
    ....111, 116-117, 122
one call closes 115
qualifying customers 103
selling yourself 32
setting job start date 103
the "new angle" 117
Seminars 211, 307
    as educational tool 21
    looking into 220
    teaching 221
Service, customer 207
Setting goals.............13-14
Shopping bids 226
Shortcuts, CD-ROM installation ..266
Showroom 41
Simon, F. J., magazine columnist ..123
Skills
    listening 97
    required, work 7
    time management 227
Sliding scale markup 67-69
    checking sales 187
Software programs
    ....193-194, 265-267, 308-309
Speaking, importance of
    grammar 209-210
Specialty contractors
    break-even chart 182
    contracting with 140-142
    gross profit ratio 187
    Overhead Expense Chart, example 37
    Overhead Expense Chart,
        See page 265
    overhead expenses 63
    relationship with other trades ...210
    Quarterly Overhead Review,
        example 51
    Quarterly Overhead Review form,
        See page 265
    sales expenses 63
    setting up an O.C.R.A. 176
    work volume 64
    working with general
        contractors 64-65
Specialty Selection Sheet, See page 265
Specifications, contract 123
Spouse, involvement in business ..224
    salary 48
Spreadsheet programs 195
Staff, resistance to change 45
Standard contract/proposal ...137-140
Start dates, determining 103
Start up
    overhead 39
    work projections 40
Startup instructions, CD-ROM 265
Statistics, failure rate 5-6, 19-20

Stick estimating 29
Stone, Michael, about the author ..317
Study, consumer buying 32
Subcontractor Change Order,
    See page 265
Subcontractor contracts 140
    sample 141-142
Subcontractors
    low bids 65-66
    thank you cards for 208
    your relationship with 210
Success
    attitude 9
    basics for 7
    hard work 9
    honesty 9
    importance of education 8, 11
    trade competency 7
    understanding markup 8
Superintendent
    training on markup 217-218
    writing work orders 153, 155
Supervision work, contract for 131-133
Suppliers, your relationship with ..210
Symptoms of failure 20-26

T

Target date, break even 182-183
Teaching in trades 7
Teaching seminars 221
Telephone
    etiquette 96, 207-208
    returning customer calls 99
Telephone book, ads in 94
Ten percent myth 34
    insurance work 203-204
    new home construction 61
Terminology, industry 26
Thank you cards 208-210
The E-Myth 7, 306
Time and materials estimating ..83-85
Time management 227
Tools and equipment, rent or buy ..185
Tools, educational 22
Total volume built, formula for 55
Tracking
    profits 186-187
    sales 160, 162
Trade associations 305
    help with contracts 128
    participation in 222
Trade magazines, as educational
    tools 22
Trades, skills needed 7-8
Training
    educational materials 22, 306
    motivational speakers 8
    on the job, trades 7
    seminars 21, 307
    video tapes 22, 306-307
Training programs
    developing 217
    employee 213, 217-218

sales 15, 89
telephone etiquette 208
time management 227
Transmittal, fax 143, 146
Travel time formula 183-184

U

Unbudgeted expenses,
    assimilating 72-77
Understanding markup 24
Uninstalling forms 266
Unit cost estimating 29
    form 30
Using forms on CD-ROM 265-267

V

Value, work 211
Variable overhead 38
Video tapes, as educational
    tools 22, 306-307
Volume, projecting for cash flow
   171-173
Volume built per employee
    commercial work 66
    new home construction 58
    residential remodeling 43
    specialty contractors 64
Volume sold, markup formula 18

W

Wages
    owner 17, 47, 185
    spouse 48
Windows™, installation of
    Markup & Profit 265-266
Wives, involvement in business ...224
Word-of-mouth advertising 63, 96
Word programs, forms on 266
Work
    calculating available 40
    owner's wage for 17
    quality of 16
    staying on budget 16-17
    value of 211
Work changes
    major 152
    minor 151
Work ethic 9
Work schedules 164, 166-167
Workers
    family as employees 48
    production per employee 44-47
Working capital, projections 24
Writing contracts 126-147
Writing letters/thank you cards 208-210
Written bids 115-116

X, Y, Z

Yellow Page ads 8, 93-94
Zig Ziglar, motivational training ..8, 9

About the Author

Michael Stone started as a "gopher" in his father's construction company in the early 1950's. He wired his first home in 1957, and installed his first forced-air heating system in 1959. He carried a plumbing license for 14 years, and worked in a variety of other building trades. He has taken more than 3,700 remodeling sales calls, and sold or worked on over 1,500 homes and commercial buildings. He has been named "National Remodeler of the Month" by the National Association of Home Builders.

Mr. Stone graduated from Eastern Oregon University with a degree in Business & Construction Management and Computer Science. His expertise lies in business management solutions and computer software for the construction industry.

He has been published in *Builder* magazine, *Qualified Remodeler*, *Professional Remodeler*, *Remodeling*, *Journal of Light Construction*, and several other national construction-oriented publications.

He currently operates *Construction Programs & Results*, a company that writes and distributes construction-related books and articles, conducts seminars for construction companies, and develops and markets software applications for residential and commercial contractors.

Frequently featured at national and regional conventions and association meetings, he speaks on business management and computer software applications for the construction industry. Since 1980, he has conducted seminars for over 17,000 contractors and governmental agencies in every major U.S. and Canadian city.

Mr. Stone is available for fee arbitration, expert witness testimony, for consulting work on construction-related issues, and for speaking engagements. Contact him at the address below:

Michael Stone
Construction Programs & Results
1001 49th Street
Washougal, WA 98671
(360) 335-1100   (888) 944-0044   Fax (360) 835-1148
e-mail: michael@markupandprofit.com
Internet: www.markupandprofit.com

Practical References for Builders

Construction Forms & Contracts

125 forms you can copy and use — or load into your computer (from the FREE disk enclosed). Then you can customize the forms to fit your company, fill them out, and print. Loads into *Word for Windows, Lotus 1-2-3, WordPerfect, Works,* or *Excel* programs. You'll find forms covering accounting, estimating, fieldwork, contracts, and general office. Each form comes with complete instructions on when to use it and how to fill it out. These forms were designed, tested and used by contractors, and will help keep your business organized, profitable and out of legal, accounting and collection troubles. Includes a CD-ROM for *Windows*™ and Mac.
400 pages, 8^1/$_2$ x 11, $41.75

CD Estimator

If your computer has *Windows*™ and a CD-ROM drive, *CD Estimator* puts at your fingertips 85,000 construction costs for new construction, remodeling, renovation & insurance repair, electrical, plumbing, HVAC and painting. You'll also have the *National Estimator* program — a stand-alone estimating program for *Windows*™ that *Remodeling* magazine called a "computer wiz." Quarterly cost updates are available at no charge on the Internet. To help you create professional-looking estimates, the disk includes over 40 construction estimating and bidding forms in a format that's perfect for nearly any word processing or spreadsheet program for *Windows*™. And to top it off, a 70-minute interactive video teaches you how to use this CD-ROM to estimate construction costs. **CD Estimator is $68.50**

National Construction Estimator

Current building costs for residential, commercial, and industrial construction. Estimated prices for every common building material. Provides man-hours, recommended crew, and gives the labor cost for installation. Includes a CD-ROM with an electronic version of the book with *National Estimator*, a stand-alone *Windows*™ estimating program, plus an interactive multimedia video that shows how to use the disk to compile construction cost estimates. **616 pages, 8^1/$_2$ x 11, $47.50. Revised annually**

Contractor's Survival Manual

How to survive hard times and succeed during the up cycles. Shows what to do when the bills can't be paid, finding money and buying time, transferring debt, and all the alternatives to bankruptcy. Explains how to build profits, avoid problems in zoning and permits, taxes, time-keeping, and payroll. Unconventional advice on how to invest in inflation, get high appraisals, trade and postpone income, and stay hip-deep in profitable work.

160 pages, 8^1/$_2$ x 11, $22.25

Steel-Frame House Construction

Framing with steel has obvious advantages over wood, yet building with steel requires new skills that can present challenges to the wood builder. This new book explains the secrets of steel framing techniques for building homes, whether pre-engineered or built stick by stick. It shows you the techniques, the tools, the materials, and how you can make it happen. Includes hundreds of photos and illustrations, plus a CD-ROM with steel framing details. **320 pages, 8^1/$_2$ x 11, $39.75**

Commercial Metal Stud Framing

Framing commercial jobs can be more lucrative than residential work. But most commercial jobs require some form of metal stud framing. This book teaches step-by-step, with hundreds of job site photos, high-speed metal stud framing in commercial construction. It describes the special tools you'll need and how to use them effectively, and the material and equipment you'll be working with. You'll find the shortcuts, tips and tricks-of-the-trade that take most steel frames years on the job to discover. Shows how to set up a crew to maintain a rhythm that will speed progress faster than any wood framing job. If you've framed with wood, this book will teach you how to be one of the few top-notch metal stud framers.
208 pages, 8^1/$_2$ x 11, $45.00

Contractor's Guide to the Building Code Revised

This new edition was written in collaboration with the International Conference of Building Officials, writers of the code. It explains in plain English exactly what the latest edition of the *Uniform Building Code* requires. Based on the 1997 code, it explains the changes and what they mean for the builder. Also covers the *Uniform Mechanical Code* and the *Uniform Plumbing Code.* Shows how to design and construct residential and light commercial buildings that'll pass inspection the first time. Suggests how to work with an inspector to minimize construction costs, what common building shortcuts are likely to be cited, and where exceptions may be granted. **320 pages, 8^1/$_2$ x 11, $39.00**

How to Succeed With Your Own Construction Business

Everything you need to start your own construction business: setting up the paperwork, finding the work, advertising, using contracts, dealing with lenders, estimating, scheduling, finding and keeping good employees, keeping the books, and coping with success. If you're considering starting your own construction business, all the knowledge, tips, and blank forms you need are here. **336 pages, 8^1/$_2$ x 11, $28.50**

Building Contractor's Exam Preparation Guide

Passing today's contractor's exams can be a major task. This book shows you how to study, how questions are likely to be worded, and the kinds of choices usually given for answers. Includes sample questions from actual state, county, and city examinations, plus a sample exam to practice on. This book isn't a substitute for the study material that your testing board recommends, but it will help prepare you for the types of questions — and their correct answers — that are likely to appear on the actual exam. Knowing how to answer these questions, as well as what to expect from the exam, can greatly increase your chances of passing.
320 pages, 8^1/$_2$ x 11, $35.00

Contractor's Index to the 1997 Uniform Building Code

Finally, there's a common-sense index that helps you quickly and easily find the section you're looking for in the *UBC*. It lists topics under the names builders actually use in construction. Best of all, it gives the full section number and the actual page in the *UBC* where you'll find it. If you need to know the requirements for windows in exit access corridor walls, just look under *Windows*™. You'll find the requirements you need are in Section 1004.3.4.3.2.2 in the *UBC* — on page 115. This practical index was written by a former builder and building inspector who knows the UBC from both perspectives. If you hate to spend valuable time hunting through pages of fine print for the information you need, this is the book for you. **192 pages, 8^1/$_2$ x 11, paperback edition, $26.00**
192 pages, 8^1/$_2$ x 11, loose-leaf edition, $29.00

Basic Engineering for Builders

If you've ever been stumped by an engineering problem on the job, yet wanted to avoid the expense of hiring a qualified engineer, you should have this book. Here you'll find engineering principles explained in non-technical language and practical methods for applying them on the job. With the help of this book you'll be able to understand engineering functions in the plans and how to meet the requirements, how to get permits issued without the help of an engineer, and anticipate requirements for concrete, steel, wood and masonry. See why you sometimes have to hire an engineer and what you can undertake yourself: surveying, concrete, lumber loads and stresses, steel, masonry, plumbing, and HVAC systems. This book is designed to help the builder save money by understanding engineering principles that you can incorporate into the jobs you bid.

400 pages, 8^1/$_2$ x 11, $36.50

Blueprint Reading for the Building Trades

How to read and understand construction documents, blueprints, and schedules. Includes layouts of structural, mechanical, HVAC and electrical drawings. Shows how to interpret sectional views, follow diagrams and schematics, and covers common problems with construction specifications. **192 pages, 5^1/$_2$ x 8^1/$_2$, $14.75**

Contractor's Guide to QuickBooks Pro 2000

This user-friendly manual walks you through QuickBooks Pro's detailed setup procedure and explains step-by-step how to create a first-rate accounting system. You'll learn in days, rather than weeks, how to use QuickBooks Pro to get your contracting business organized, with simple, fast accounting procedures. On the CD included with the book you'll find a QuickBooks Pro file preconfigured for a construction company (you drag it over onto your computer and plug in your own company's data). You'll also get a complete estimating program, including a database, and a job costing program that lets you export your estimates to QuickBooks Pro. It even includes many useful construction forms to use in your business.
304 pages, 8½ x 11, $44.50

Basic Lumber Engineering for Builders

Beam and lumber requirements for many jobs aren't always clear, especially with changing building codes and lumber products. Most of the time you rely on your own "rules of thumb" when figuring spans or lumber engineering. This book can help you fill the gap between what you can find in the building code span tables and what you need to pay a certified engineer to do. With its large, clear illustrations and examples, this book shows you how to figure stresses for pre-engineered wood or wood structural members, how to calculate loads, and how to design your own girders, joists and beams. Included FREE with the book — an easy-to-use limited version of NorthBridge Software's *Wood Beam Sizing* program.
272 pages, 8½ x 11, $38.00

Electrician's Exam Preparation Guide

Need help in passing the apprentice, journeyman, or master electrician's exam? This is a book of questions and answers based on actual electrician's exams over the last few years. Almost a thousand multiple-choice questions — exactly the type you'll find on the exam — cover every area of electrical installation: electrical drawings, services and systems, transformers, capacitors, distribution equipment, branch circuits, feeders, calculations, measuring and testing, and more. It gives you the correct answer, an explanation, and where to find it in the latest *NEC*. Also tells how to apply for the test, how best to study, and what to expect on examination day.
352 pages, 8½ x 11, $32.00

Craftsman's Illustrated Dictionary of Construction Terms

Almost everything you could possibly want to know about any word or technique in construction. Hundreds of up-to-date construction terms, materials, drawings and pictures with detailed, illustrated articles describing equipment and methods. Terms and techniques are explained or illustrated in vivid detail. Use this valuable reference to check spelling, find clear, concise definitions of construction terms used on plans and construction documents, or learn about little-known tools, equipment, tests and methods used in the building industry. It's all here.
416 pages, 8½ x 11, $36.00

Scheduling With Microsoft Project 2000

Step-by-step instructions for using this software to keep your projects on schedule and within budget. Learn to adjust time scales, milestones, and tasks, assign and track resources and costs, monitor and update the schedule with baselines, keep track of changes to the schedule, compare and analyze actual costs, and record expenditures. With this powerful tool, you'll not only be able to track a job's progress, you can make reports of exactly where on the timeline everything is and when it will be done. Includes a disk with examples from a typical residential schedule. Requires that you have *Microsoft Project 2000* on your computer.
128 pages, 7 x 9, $49.95

Contracting in All 50 States

Every state has its own licensing requirements that you must meet to do business there. These are usually written exams, financial requirements, and letters of reference. This book shows how to get a building, mechanical or specialty contractor's license, qualify for DOT work, and register as an out-of-state corporation, for every state in the U.S. It lists addresses, phone numbers, application fees, requirements, where an exam is required, what's covered on the exam and how much weight each area of construction is given on the exam. You'll find just about everything you need to know in order to apply for your out-of-state license.
416 pages, 8½ x 11, $36.00

Illustrated Guide to the 1999 National Electrical Code

This fully-illustrated guide offers a quick and easy visual reference for installing electrical systems. Whether you're installing a new system or repairing an old one, you'll appreciate the simple explanations written by a code expert, and the detailed, intricately-drawn and labeled diagrams. A real time-saver when it comes to deciphering the current *NEC*.
360 pages, 8½ x 11, $38.75

CD Estimator — Heavy

CD Estimator — Heavy has a complete 780-page heavy construction cost estimating volume for each of the 50 states. Select the cost database for the state where the work will be done. Includes thousands of cost estimates you won't find anywhere else, and in-depth coverage of demolition, hazardous materials remediation, tunneling, site utilities, precast concrete, structural framing, heavy timber construction, membrane waterproofing, industrial windows and doors, specialty finishes, built-in commercial and industrial equipment, and HVAC and electrical systems for commercial and industrial buildings. **CD Estimator — Heavy is $69.00**

The Contractor's Legal Kit

Stop "eating" the costs of bad designs, hidden conditions, and job surprises. Set ground rules that assign those costs to the rightful party ahead of time. And it's all in plain English, not "legalese." For less than the cost of an hour with a lawyer you'll learn the exclusions to put in your agreements, why your insurance company may pay for your legal defense, how to avoid liability for injuries to your sub and his employees or damages they cause, how to collect on lawsuits you win, and much more. It also includes a FREE computer disk with contracts and forms you can customize for your own use.
352 pages, 8½ x 11, $59.95

Contractor's Year-Round Tax Guide Revised

How to set up and run your construction business to minimize taxes: corporate tax strategy and how to use it to your advantage, and what you should be aware of in contracts with others. Covers tax shelters for builders, write-offs and investments that will reduce your taxes, accounting methods that are best for contractors, and what the I.R.S. allows and what it often questions. **192 pages, 8½ x 11, $26.50**

Construction Estimating Reference Data

Provides the 300 most useful manhour tables for practically every item of construction. Labor requirements are listed for sitework, concrete work, masonry, steel, carpentry, thermal and moisture protection, doors and windows, finishes, mechanical and electrical. Each section details the work being estimated and gives appropriate crew size and equipment needed. Includes a CD-ROM with an electronic version of the book with *National Estimator*, a stand-alone *Windows*™ estimating program, plus an interactive multimedia video that shows how to use the disk to compile construction cost estimates. **432 pages, 11 x 8½, $39.50**

Estimating Excavation

How to calculate the amount of dirt you'll have to move and the cost of owning and operating the machines you'll do it with. Detailed, step-by-step instructions on how to assign bid prices to each part of the job, including labor and equipment costs. Also, the best ways to set up an organized and logical estimating system, take off from contour maps, estimate quantities in irregular areas, and figure your overhead. **448 pages, 8½ x 11, $39.50**

Rough Framing Carpentry

If you'd like to make good money working outdoors as a framer, this is the book for you. Here you'll find shortcuts to laying out studs; speed cutting blocks, trimmers and plates by eye; quickly building and blocking rake walls; installing ceiling backing, ceiling joists, and truss joists; cutting and assembling hip trusses and California fills; arches and drop ceilings — all with production line procedures that save you time and help you make more money. Over 100 on-the-job photos of how to do it right and what can go wrong. **304 pages, 8½ x 11, $26.50**

Home Inspection Handbook

Every area you need to check in a home inspection — especially in older homes. Twenty complete inspection checklists: building site, foundation and basement, structural, bathrooms, chimneys and flues, ceilings, interior & exterior finishes, electrical, plumbing, HVAC, insects, vermin and decay, and more. Also includes information on starting and running your own home inspection business. **324 pages, 5^{1}/$_{2}$ x 8^{1}/$_{2}$, $24.95**

Estimating Home Building Costs

Estimate every phase of residential construction from site costs to the profit margin you include in your bid. Shows how to keep track of man-hours and make accurate labor cost estimates for footings, foundations, framing and sheathing finishes, electrical, plumbing, and more. Provides and explains sample cost estimate worksheets with complete instructions for each job phase. **320 pages, 5^{1}/$_{2}$ x 8^{1}/$_{2}$, $17.00**

Land Development

The industry's bible. Nine chapters cover everything you need to know about land development from initial market studies to site selection and analysis. New and innovative design ideas for streets, houses, and neighborhoods are included. Whether you're developing a whole neighborhood or just one site, you shouldn't be without this essential reference. **360 pages, 5^{1}/$_{2}$ x 8^{1}/$_{2}$, $62.00**

Estimating With Microsoft Excel

Most builders estimate with Excel because it's easy to learn, quick to use, and can be customized to your style of estimating. Here you'll find step-by-step instructions on how to create your own customized automated spreadsheet estimating program for use with Excel. You'll learn how to use the magic of Excel to create detail sheets, cost breakdown summaries, and links. You'll put this all to use in estimating concrete, rebar, permit fees, and roofing. You can even create your own macros. Includes a CD-ROM that illustrates examples in the book and provides you with templates you can use to set up your own estimating system. **150 pages, 7 x 9, $49.95**

Getting Financing & Developing Land

Developing land is a major leap for most builders - y that's where the big money is made. This book gives yo the practical knowledge you need to make that lea Learn how to prepare a market study, select a buildin site, obtain financing, guide your plans through approva then control your building costs so you can ensure you self a good profit. Includes a CD-ROM with forms, chec lists, and a sample business plan you can customize an use to help you sell your idea to lenders and investors. **232 pages, 8^{1}/$_{2}$ x 11, $39.00**

Moving to Commercial Construction

In commercial work, a single job can keep you and your crews busy for year or more. The profit percentages are higher, but so is the risk involve This book takes you step-by-step through the process of setting up a suc cessful commercial business; finding work, estimating and bidding, valu engineering, getting through the submittal and shop drawing proces keeping a stable work force, controlling costs, and promoting your bus ness. Explains the design/build and partnering business concepts an their advantage over the competitive bid process. Includes sample letter contracts, checklists and forms that you can use in your business, plus CD-ROM with blank copies in several word-processing formats. **256 pages, 8^{1}/$_{2}$ x 11, $42.00**

Builder's Guide to Accounting Revised

Step-by-step, easy -to-follow guidelines for setting up and maintaining records for your building business. This practical, newly-revised guide to all accounting methods shows how to meet state and federal accounting requirements, explains the new depreciation rules, and describes how the Tax Reform Act can affect the way you keep records. Full of charts, diagrams, simple directions and examples, to help you keep track of where your money is going. Recommended reading for many state contractor's exams. **320 pages, 8^{1}/$_{2}$ x 11, $26.50**

Craftsman Book Company
6058 Corte del Cedro
P.O. Box 6500
Carlsbad, CA 92018

☎ **24 hour order line**
1-800-829-8123
Fax (760) 438-0398

Name _____
e-mail address (for order tracking and special offers) _____
Company _____
Address _____
City/State/Zip _____
○ This is a residence

Total enclosed_____(In California add 7% tax)
We pay shipping when your check covers your order in full.

In A Hurry?
We accept phone orders charged to your
○ Visa, ○ MasterCard, ○ Discover or ○ American Express

Card#_____
Exp. date_____Initials_____

Tax Deductible: Treasury regulations make these references tax deductible when used in your work. Save the canceled check or charge card statement as your receipt.

Order online http://www.craftsman-book.com
Free on the Internet! Download any of Craftsman's estimating costbooks for a 30-day free trial! http://costbook.com

10-Day Money Back Guarantee

○ 36.50 Basic Engineering for Builders
○ 38.00 Basic Lumber Engineering for Builders
○ 14.75 Blueprint Reading for the Building Trades
○ 26.50 Builder's Guide to Accounting Revised
○ 35.00 Building Contractor's Exam Preparation Guide
○ 68.50 CD Estimator
○ 69.00 CD Estimator Heavy
○ 45.00 Commercial Metal Stud Framing
○ 39.50 Construction Estimating Reference Data with FREE National Estimator on a CD-ROM.
○ 41.75 Construction Forms & Contracts with a CD-ROM for WindowsTM and Macintosh.
○ 36.00 Contracting in All 50 States
○ 44.50 Contractor's Guide to QuickBooks Pro 2000
○ 39.00 Contractor's Guide to the Building Code Revised
○ 26.00 Contractor's Index to the UBC - Paperback
○ 29.00 Contractor's Index to the UBC - Loose-leaf
○ 59.95 Contractor's Legal Kit
○ 22.25 Contractor's Survival Manual
○ 26.50 Contractor's Year-Round Tax Guide Revised
○ 36.00 Craftsman's Illustrated Dictionary of Construction Terms
○ 32.00 Electrician's Exam Preparation Guide
○ 39.50 Estimating Excavation
○ 17.00 Estimating Home Building Costs
○ 49.95 Estimating With Microsoft Excel
○ 39.00 Getting Financing & Developing Land
○ 24.95 Home Inspection Handbook
○ 28.50 How to Succeed w/Your Own Construction Business
○ 38.75 Illustrated Guide to the 1999 National Electrical Code
○ 62.00 Land Development
○ 42.00 Moving to Commercial Construction
○ 47.50 National Construction Estimator with FREE National Estimator on a CD-ROM.
○ 26.50 Rough Framing Carpentry
○ 49.95 Scheduling with Microsoft Project 2000
○ 39.75 Steel-Frame House Construction
○ 32.50 Markup & Profit: A Contractor's Guide
○ FREE Full Color Catalog

Prices subject to change without notice

From the Author of "Markup and Profit"...

More Help for Contractors! Please check off any items of interest. Then, return this card or contact us at the phone number or web address below.

Send us your e-mail address and we'll put you on the list to receive our FREE newsletter!

- ❏ Audio & Video Tapes Based on This Book
- ❏ Consulting Services
- ❏ Speaking Engagements
- ❏ Marketing and Promotion Aids
- ❏ Computer Hardware/Software
- ❏ Forms & Contracts

Don't Delay Reply Today!

Name_____
Company_____
Address_____
City/State/Zip_____
e-mail_____
Phone_____ Fax_____

Return card to: Construction Programs & Results
1001 49th St. • Washougal, WA 98671

1-888-944-0044 / www.markupandprofit.com

 Craftsman Book Company
6058 Corte del Cedro
P.O. Box 6500
Carlsbad, CA 92018

☎ **24 hour order line**
1-800-829-8123
Fax (760) 438-0398

In A Hurry?
We accept phone orders charged to your
○ Visa, ○ MasterCard, ○ Discover or ○ American Express

Total enclosed_____ (In California add 7% tax)
We pay shipping when your check covers your order in full.

Card#_____
Exp. date_____ Initials_____

Name_____
e-mail address (for order tracking and special offers)_____
Company_____
Address_____
City/State/Zip_____ ○ This is a residence

Tax Deductible: Treasury regulations make these references tax deductible when used in your work. Save the canceled check or charge card statement as your receipt.

Order online http://www.craftsman-book.com
Free on the Internet! Download any of Craftsman's estimating costbooks for a 30-day free trial! http://costbook.com

10-Day Money Back Guarantee

- ○ 36.50 Basic Engineering for Builders
- ○ 38.00 Basic Lumber Engineering for Builders
- ○ 14.75 Blueprint Reading for the Building Trades
- ○ 26.50 Builder's Guide to Accounting Revised
- ○ 35.00 Building Contractor's Exam Preparation Guide
- ○ 68.50 CD Estimator
- ○ 69.00 CD Estimator Heavy
- ○ 45.00 Commercial Metal Stud Framing
- ○ 39.50 Construction Estimating Reference Data with FREE National Estimator on a CD-ROM.
- ○ 41.75 Construction Forms & Contracts with a CD-ROM for Windows™ and Macintosh.
- ○ 36.00 Contracting in All 50 States
- ○ 44.50 Contractor's Guide to QuickBooks Pro 2000
- ○ 39.00 Contractor's Guide to the Building Code Revised
- ○ 26.00 Contractor's Index to the UBC - Paperback
- ○ 29.00 Contractor's Index to the UBC - Loose-leaf
- ○ 59.95 Contractor's Legal Kit
- ○ 22.25 Contractor's Survival Manual
- ○ 26.50 Contractor's Year-Round Tax Guide Revised
- ○ 36.00 Craftsman's Illustrated Dictionary of Constr. Terms
- ○ 32.00 Electrician's Exam Preparation Guide
- ○ 39.50 Estimating Excavation
- ○ 17.00 Estimating Home Building Costs
- ○ 49.95 Estimating With Microsoft Excel
- ○ 39.00 Getting Financing & Developing Land
- ○ 24.95 Home Inspection Handbook
- ○ 28.50 How to Succeed w/Your Own Construction Business
- ○ 38.75 Illustrated Guide to the 1999 Natl Electrical Code
- ○ 62.00 Land Development
- ○ 42.00 Moving to Commercial Construction
- ○ 47.50 National Construction Estimator with FREE National Estimator on a CD-ROM.
- ○ 26.50 Rough Framing Carpentry
- ○ 49.95 Scheduling with Microsoft Project 2000
- ○ 39.75 Steel-Frame House Construction
- ○ 32.50 Markup & Profit: A Contractor's Guide
- ○ FREE Full Color Catalog

Prices subject to change without notice

Mail This Card Today
For a Free Full Color Catalog

Over 100 books, annual cost guides and estimating software packages at your fingertips with information that can save you time and money. Here you'll find information on carpentry, contracting, estimating, remodeling, electrical work, and plumbing.

All items come with an unconditional 10-day money-back guarantee. If they don't save you money, mail them back for a full refund.

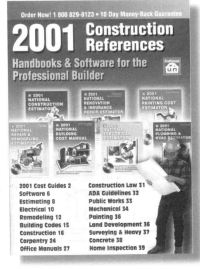

Name_____
e-mail address (for special offers)_____
Company_____
Address_____
City/State/Zip_____

Craftsman Book Company / 6058 Corte del Cedro / P.O. Box 6500 / Carlsbad, CA 92018

Construction Programs & Results
1001 49th Street
Washougal, WA 98671

Please Affix $.34 Postage Here

BUSINESS REPLY MAIL
FIRST CLASS MAIL PERMIT NO. 271 CARLSBAD, CA
POSTAGE WILL BE PAID BY ADDRESSEE

Craftsman Book Company
6058 Corte del Cedro
P.O. Box 6500
Carlsbad, CA 92018-9974

NO POSTAGE NECESSARY IF MAILED IN THE UNITED STATES

BUSINESS REPLY MAIL
FIRST CLASS MAIL PERMIT NO. 271 CARLSBAD, CA
POSTAGE WILL BE PAID BY ADDRESSEE

Craftsman Book Company
6058 Corte del Cedro
P.O. Box 6500
Carlsbad, CA 92018-9974

NO POSTAGE NECESSARY IF MAILED IN THE UNITED STATES

Nikola Tesla and The ¨369
Decoding God's Thumbpri

Nikola Tesla did countless mysterious experiments, but he was a whole other mystery on his own.

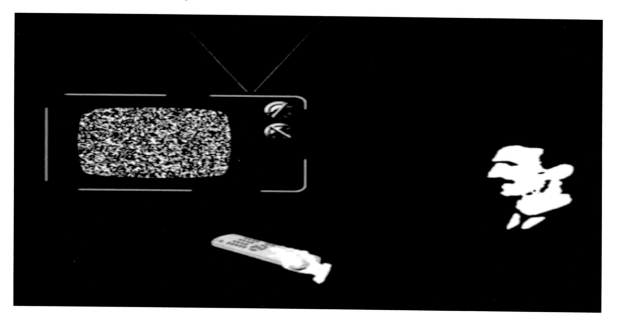

Almost all genius Minds have a certain obsession, Nikola Tesla had a pretty big one.

Nikola Tesla and The ¨369 Code¨ / Decoding God's Thumbprint
By; *Leonardo Barrios Beretta*

Tesla had some interesting habits that he would do, for example he would walk on the block repeatedly three times before entering a building.

He would clean his plates with 18 napkins.

He lived in hotel rooms only with the number 8 divisible by three.

He would make calculations about things and his immediate environment to make sure the result is divisible by three.

He would do everything in sets of 3, some say he had OCD and others say he was very superstitious, however the truth is a lot deeper.

His obsession was not simply with numbers, but especially these numbers 3, 6 and 9.

Tesla claimed that these numbers were extremely important and nobody listened, ¿But why these numbers? ¿What did Nikola Tesla tried to make the world understand?

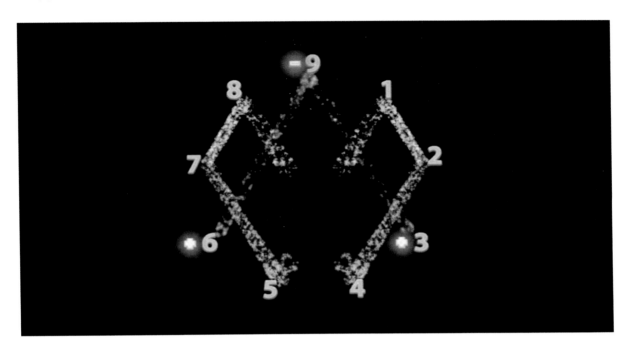

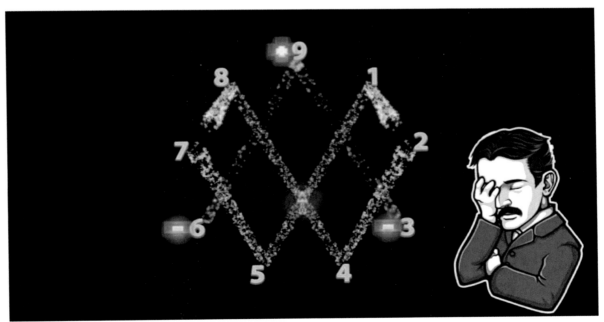

First we must understand that we did not create math, we discovered it.

Nikola Tesla and The "369 Code" / Decoding God's Thumbprint
By; *Leonardo Barrios Beretta*

It's the universal language in law, no matter where you were in the universe 1 plus 2 were always equal to 3.

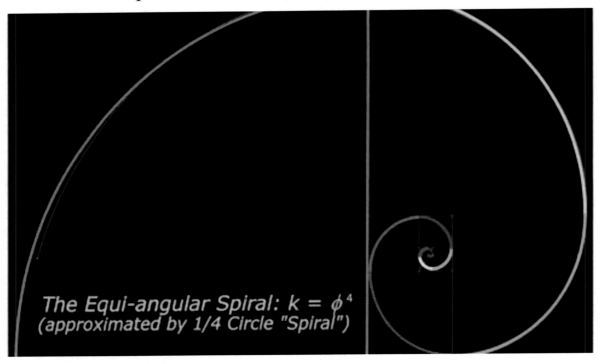

Everything in the universe will be his law, there are patterns that naturally occur in the universe.

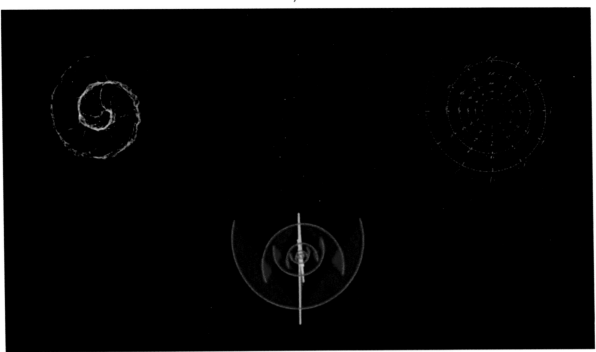

Patterns that we discovered in life, galaxies, star formations, evolution and almost all natural systems, some of these patterns are the golden ratio and sacred geometry, one really important system that nature seems to obey is the power up to binary system.

In which the patterns start from one and continues doubling the numbers.

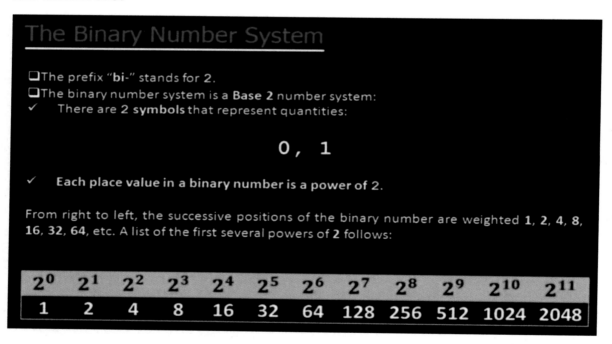

Cells and embryos develop following this sacred pattern 1, 2, 4, 8, 16, 32, etcetera.

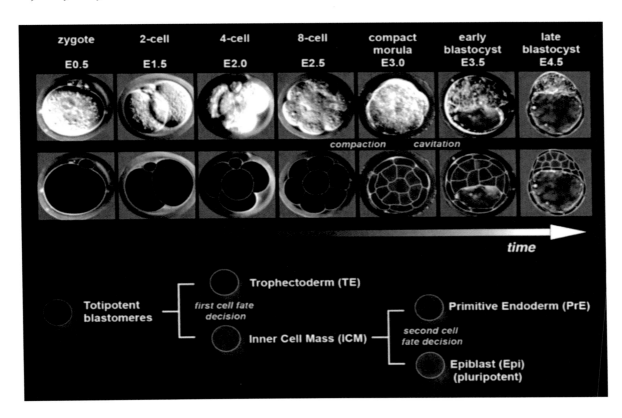

Some Colley's patterns is the blueprint of God math.

By this analogy would be God's thumbprint leaving all religion aside in to a vortex math.

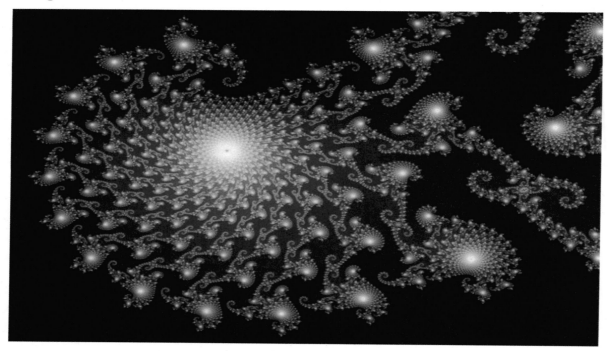

The science of Taurus Anatomy, there is a pattern that repeats itself, 1, 2, 4, 8, 7 and 5.

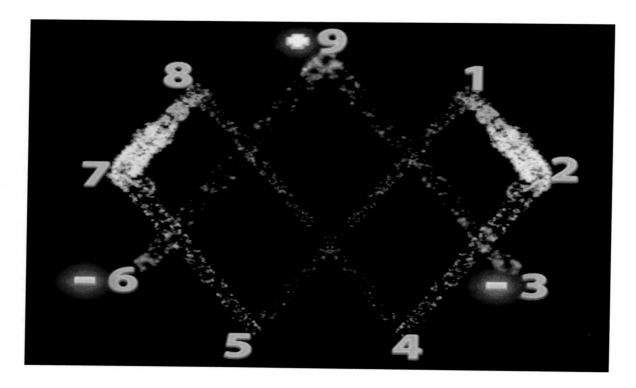

And so on as you can see 3, 6 and 9 are not in this pattern.

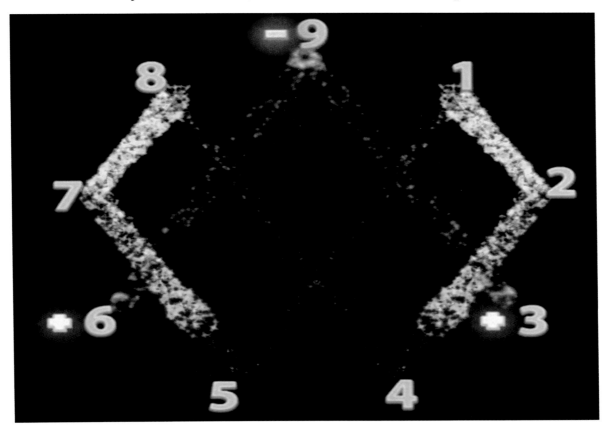

Scientist Marco Rawdon believes these numbers represent a vector from the third and fourth dimension.

Which he calls a flux field, this field is supposed to be a higher dimensional energy that influences the energy circuit of the other six points.

Randy Powell, a student of Marco Rawdon, says that this is the secret key to free energy, something we all know Tesla mastered.

Let me explain, let's start from one.

$$1+1=2$$

$$2+2=4$$

$$4+4=8$$

$$8+8=16$$

Which means that;

$$1+6=7$$

$$16+16=32$$

Resulting that;

$$3+2=5$$

You can do 7 doubled if you want to, which you would get 14.

$$32+32=64$$
$$6+4=10$$
$$1+0=1$$

So if we continue, we will keep following the same pattern.

1, 2, 4, 8, 7, 5

1, 2, 4, 8, 7, 5

1, 2, 4, 8, 7, 5

1, 2, 4, 8, 7, 5

1, 2, 4, 8, 7, 5

As you can see there is no mention of 3, 6, and 9, it's like they are beyond this pattern free from it.

However, there is something strange once you start doubling them.

3 + 3 = 6

6 + 6 = 12

Which would result in;

1 + 2 = 3

In this pattern there is no mention of ¨ 9 ¨, It's like ¨ 9 ¨ is beyond completely free from both patterns.

$$12+12=24$$
$$2+4=6$$

$$24+24=48$$
$$4+8=12$$
$$1+2=3$$

$$48+48=96$$
$$9+6=15$$
$$1+5=6$$

$$96+96=192$$
$$1+9+2=12$$
$$1+2=3$$

$$192 + 192 = 384$$
$$3 + 8 + 4 = 15$$
$$1 + 5 = 6$$

$$384 + 384 = 768$$
$$7 + 6 + 8 = 21$$
$$2 + 1 = 3$$

$$768 + 768 = 1536$$
$$1 + 5 + 3 + 6 = 15$$
$$1 + 5 = 6$$

But if you start a ¨ 9 ¨, it will always result in ¨ 9 ¨.

$$9 + 9 = 18$$
$$1 + 8 = 9$$

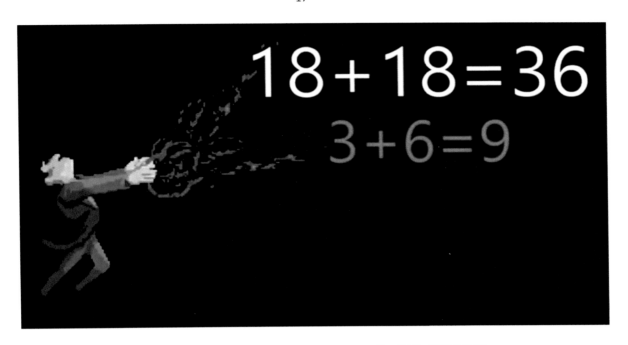

Nikola Tesla and The "369 Code" / Decoding God's Thumbprint
By; *Leonardo Barrios Beretta*

$$288+288=576$$
$$5+7+6=18$$
$$1+8=9$$

$$576+576=1152$$
$$1+1+5+2=9$$

$$1152+1152=2304$$
$$2+3+0+4=9$$

Nikola Tesla and The ¨369 Code¨ / Decoding God's Thumbprint
By; *Leonardo Barrios Beretta*

This is called the symbol of enlightenment.

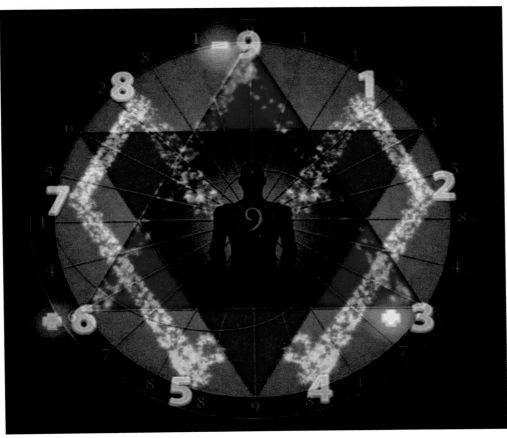

Nikola Tesla and The "369 Code" / Decoding God's Thumbprint
By; *Leonardo Barrios Beretta*

If we go to the Great Pyramid of Giza, not only are there three larger Pyramids all side by side.

Mirroring the positions of the stars in Orion's belt.

Nikola Tesla and The ¨369 Code¨ / Decoding God's Thumbprint
By; *Leonardo Barrios Beretta*

But we also see a group of three smaller pyramids immediately away from the three larger pyramids.

We find lots of evidence, that nature uses three-fold and six-fold symmetry.

"Where there is matter, there is geometry."
~ Johannes Kepler

These shapes are in nature.

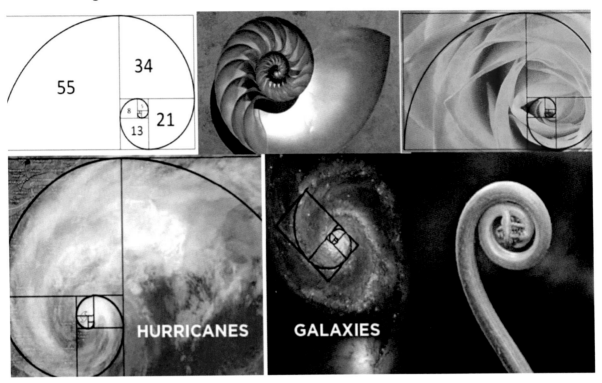

And the ancients emulated these shapes in the buildings of their sacred architecture.

Nikola Tesla and The ¨369 Code¨ / Decoding God's Thumbprint
By; *Leonardo Barrios Beretta*

Nikola Tesla and The ¨369 Code¨ / Decoding God's Thumbprint
By; *Leonardo Barrios Beretta*

¿Is it possible that Tesla uncovered this profound secret and uses knowledge to push the boundaries of Science and Technology?

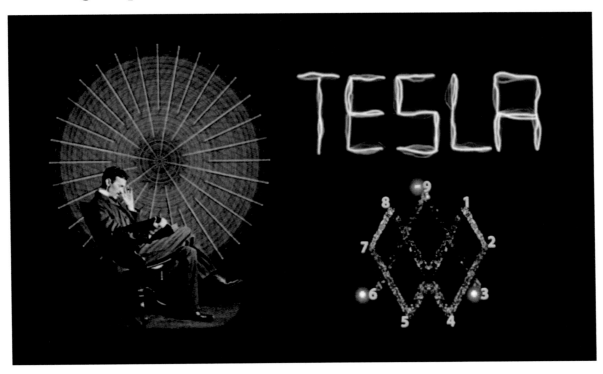

Nikola Tesla and The ¨369 Code¨ / Decoding God's Thumbprint
By; *Leonardo Barrios Beretta*

The Magnificence of 9

Let's say there are two opposites, you can call them white and black if you want.

They are like the north and south poles of a magnet, one side is 1, 2 and 4.

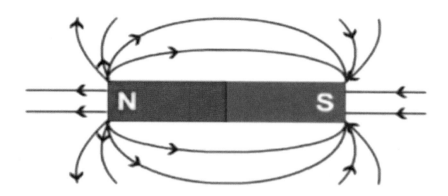

The other side is 8, 7 and 5, just like electricity, everything in the universe is a stream between these two polar sites.

Like a swinging pendulum and if you could imagine the movement it's something like the symbol for infinity.

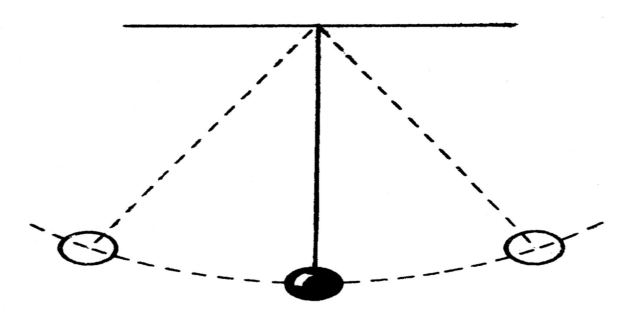

However, these suicides are governed by 3 and 6.

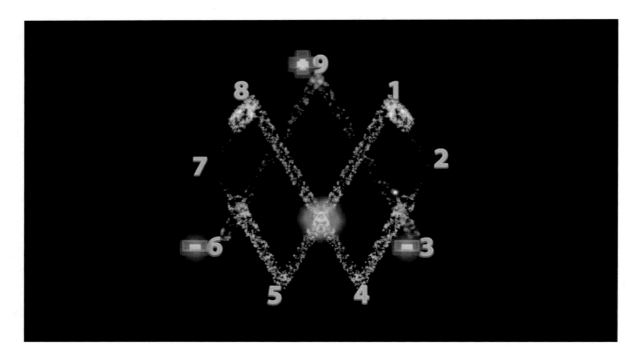

3 governs 1, 2 and 4.

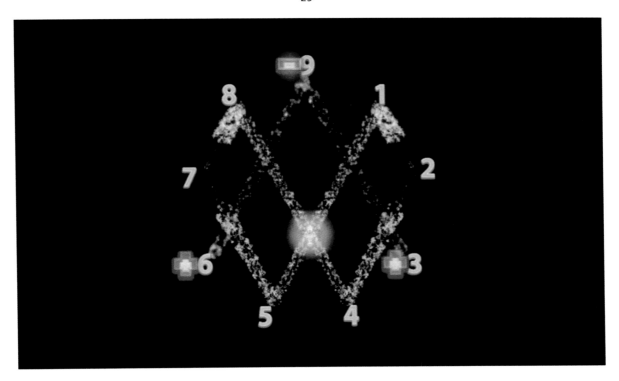

While 6 governs 8, 7 and 5.

And if you look at the pattern closely, it gets even more mind-boggling...

1 and 2 equals 3, 2 and 4 equals 6, 4 and 8 equals 3, 8 and 7 equals 6, 7 and 5 equals 3, 5 and 1 equals 6 and 1 in 2 equals 3.

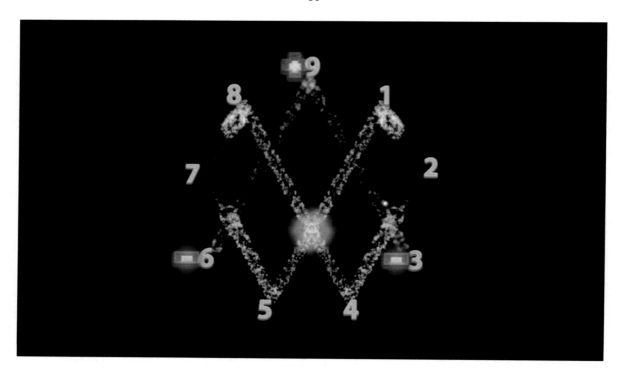

The same pattern on a higher scale is actually **3, 6, 3, 6, 3, 3.**

But even these **2, 3** and **6** are governed by **9**, which shows something spectacular.

Looking closely at the pattern of **3** and **6** you realize that **3 + 6** equals **9** and **6 + 3** equals **9**.

All the numbers together equal **9** both ways excluding and including **3 + 6.**

So ¨ **9** ¨ means the unity of the both sides, ¨ **9** ¨ is the universe itself, the vibration, the energy.

And the frequency **3, 6** and **9**...

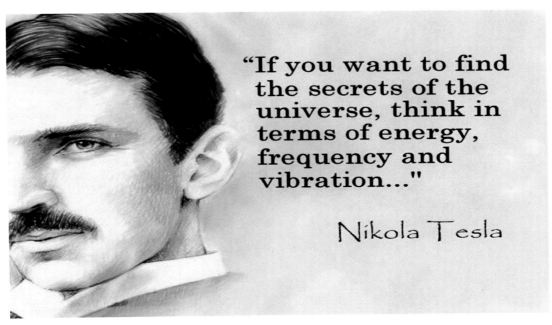

There is a deeper philosophical truth in this, let's imagine what we could accomplish if we could apply the sacred knowledge in everyday science.

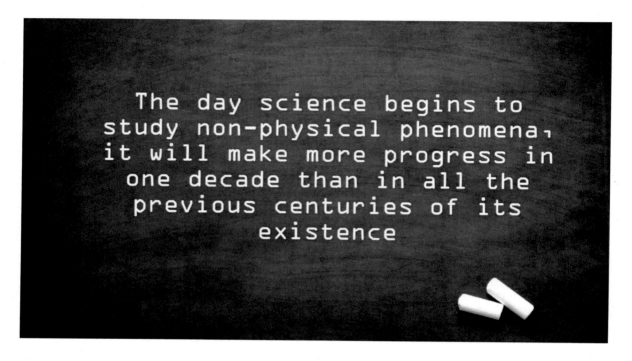

Nikola Tesla;

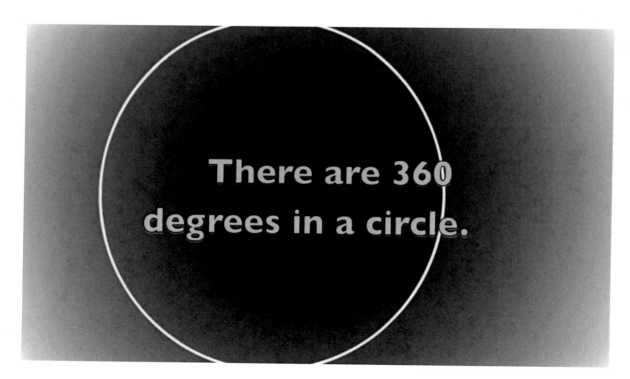

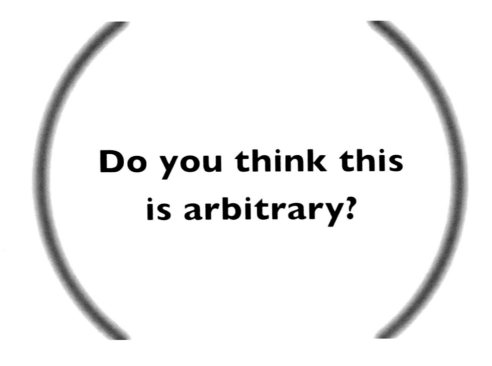

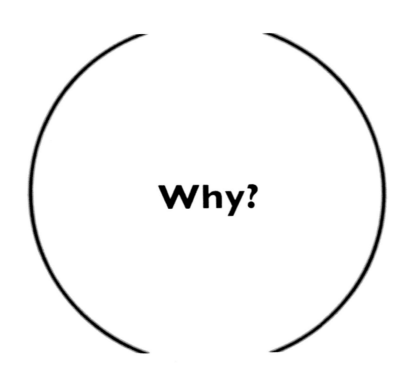

Nikola Tesla and The "369 Code" / Decoding God's Thumbprint
By; *Leonardo Barrios Beretta*

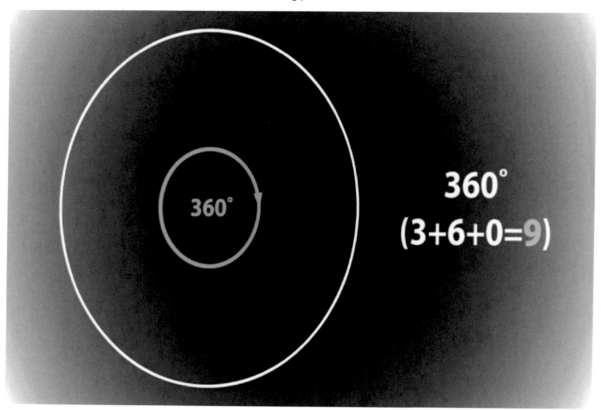

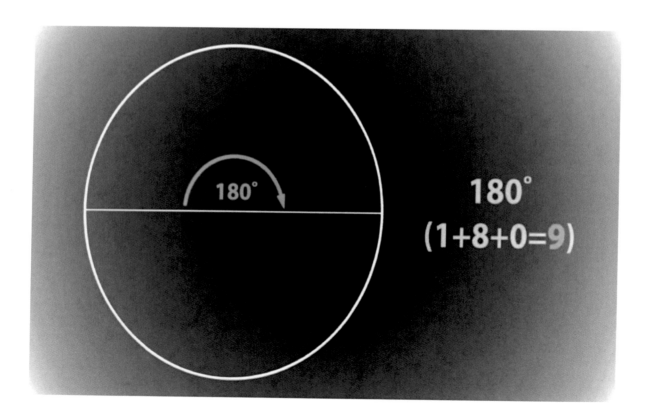

Nikola Tesla and The ¨369 Code¨ / Decoding God's Thumbprint
By; *Leonardo Barrios Beretta*

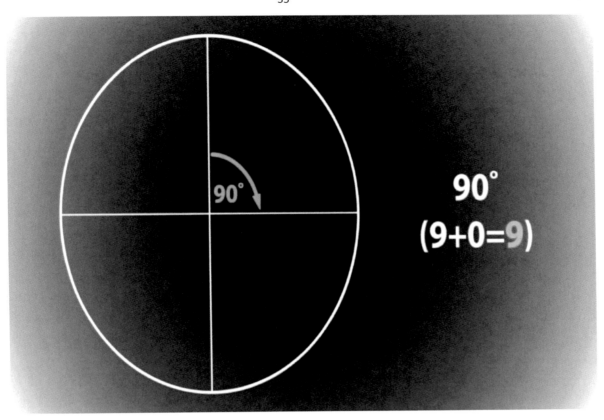

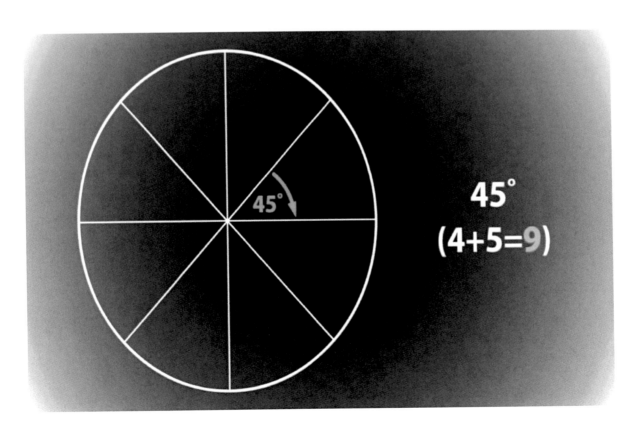

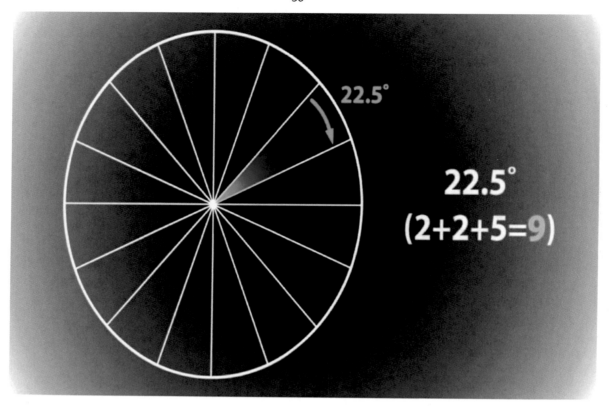

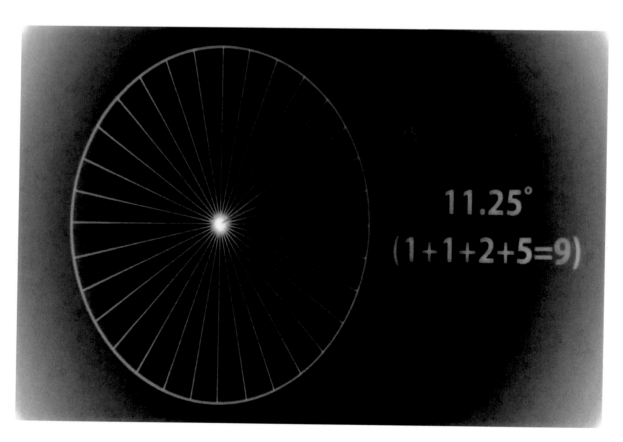

Nikola Tesla and The ¨369 Code¨ / Decoding God's Thumbprint
By; *Leonardo Barrios Beretta*

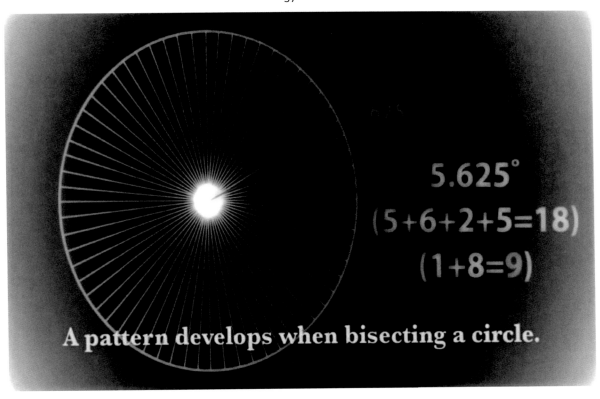

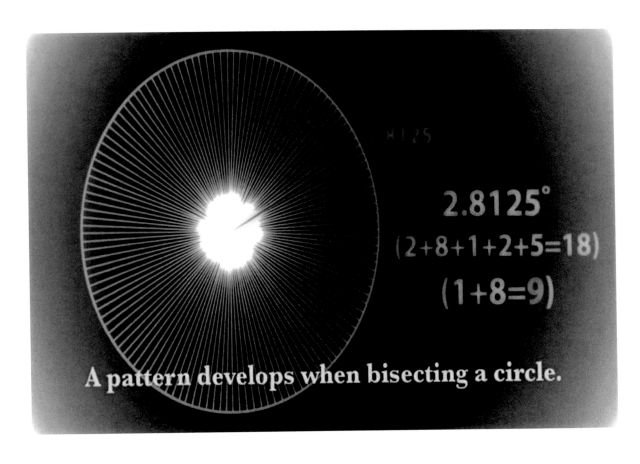

Nikola Tesla and The ¨369 Code¨ / Decoding God's Thumbprint
By; *Leonardo Barrios Beretta*

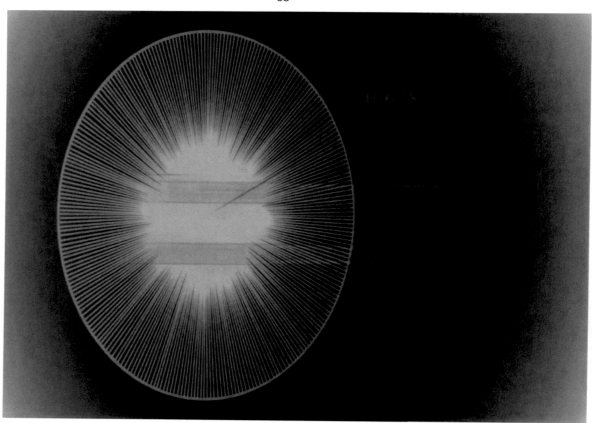

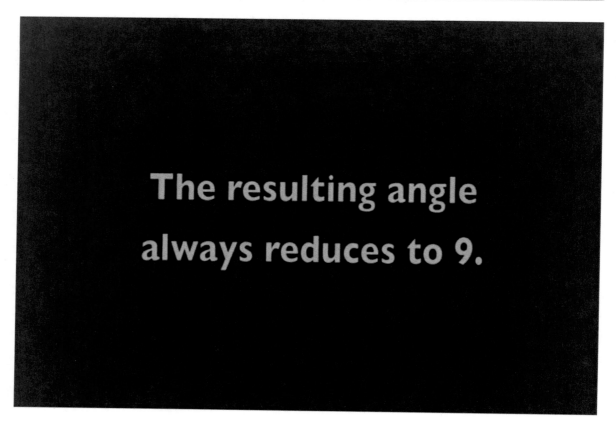

Nikola Tesla and The ¨369 Code¨ / Decoding God's Thumbprint
By; *Leonardo Barrios Beretta*

¿Is there a Divine code embedded in our numbers system?

Vortex Based Mathematics says, yes..

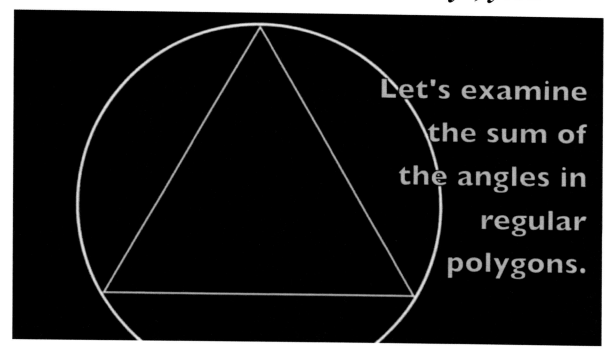

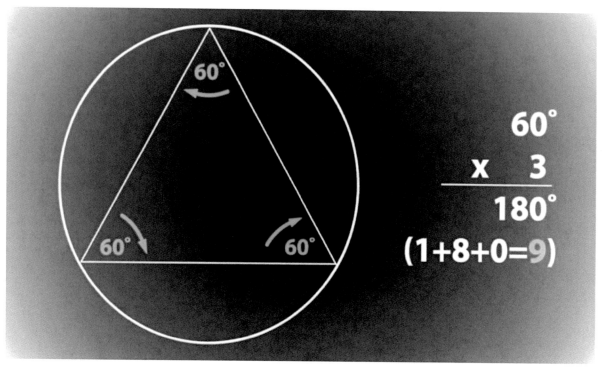

Nikola Tesla and The ¨369 Code¨ / Decoding God's Thumbprint
By; *Leonardo Barrios Beretta*

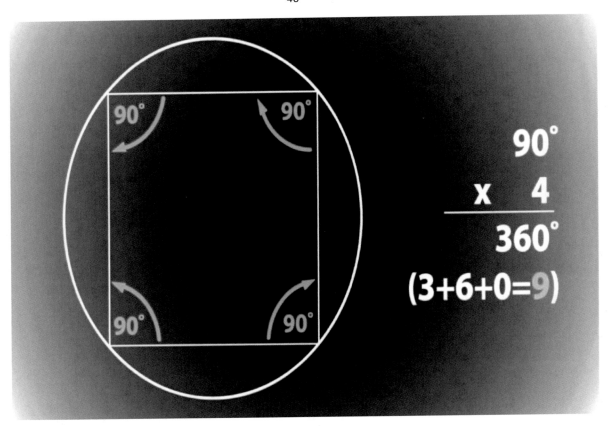

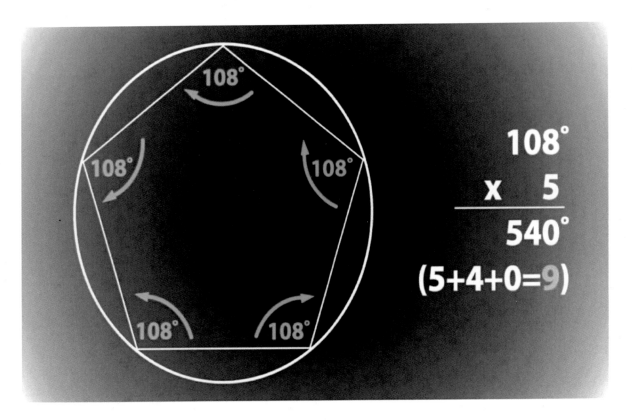

Nikola Tesla and The ¨369 Code¨ / Decoding God's Thumbprint
By; *Leonardo Barrios Beretta*

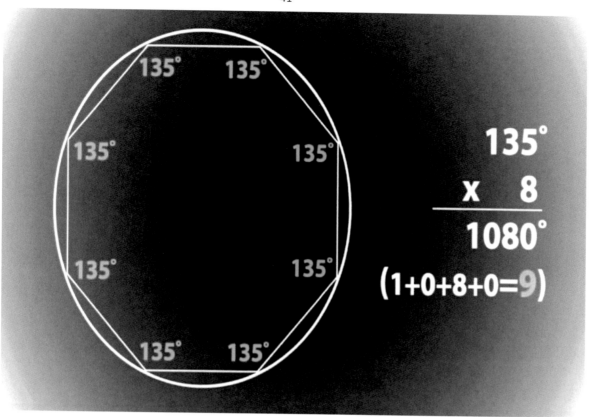

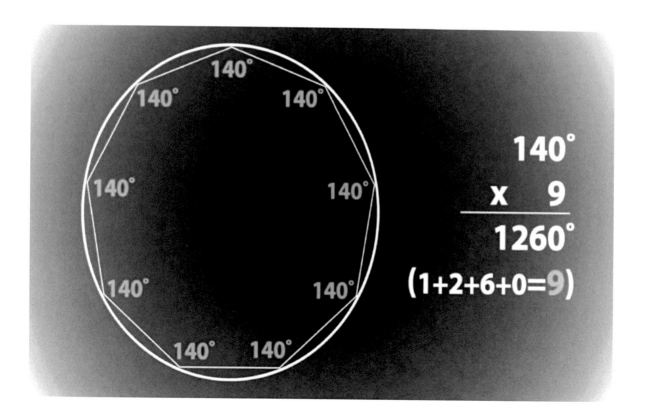

Nikola Tesla and The ¨369 Code¨ / Decoding God's Thumbprint
By; *Leonardo Barrios Beretta*

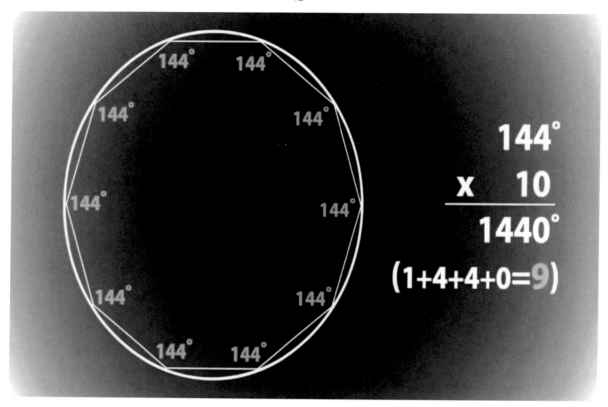

¿Meaningless Numerology?
¿Or divine symmetrology?

the resulting angle always reduces to nine.

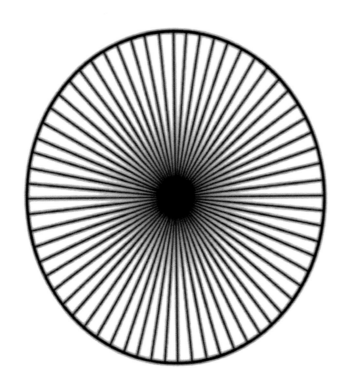

Converging into a singularity.

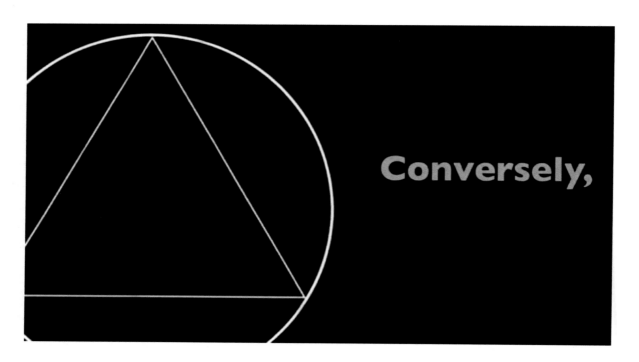

Conversely,

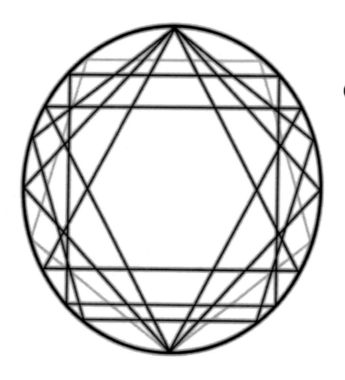

Our polygons revealed the the exact opposite.

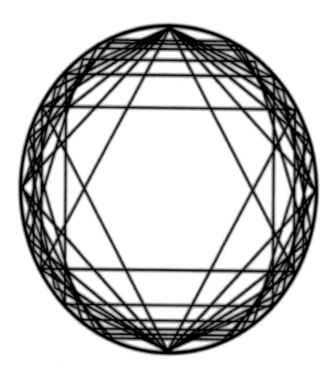

Their vectors communicate an outward divergence.

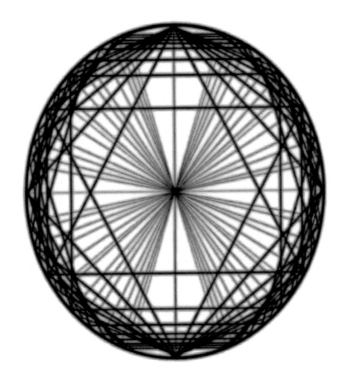

The nine reveals a linear duality.

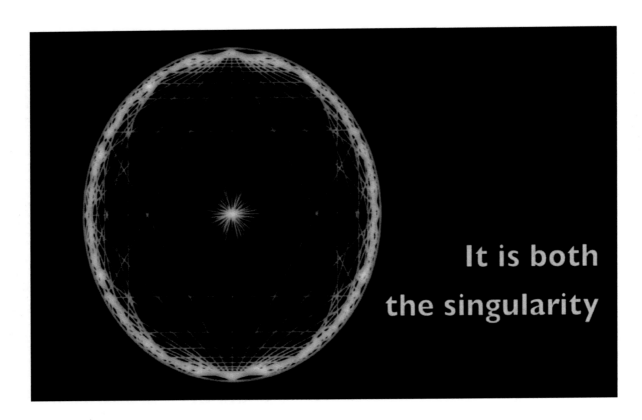

It is both the singularity

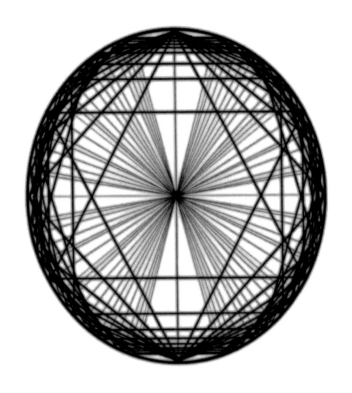

and the vacuum.

Nine models ¨Everything¨ and ¨Nothing¨ simultaneously.

¿What do I mean by that?

The sum of all digits excluding nine is 36.
1+2+3+4+5+6+7+8=36
(3+6=9)

Paradoxically, Nine plus any digit returns the same digit. i.e. 9+5=14 (1+4=5)

So nine quite literally equals all the digits (36) and nothing (0).

-360-

So, ¿Why was 6 Afraid of 7?

¡Because 7, 8, 9!

Thank you for reading...

Nikola Tesla and The ¨369 Code¨ / Decoding God's Thumbprint
By; *Leonardo Barrios Beretta*

The most read...

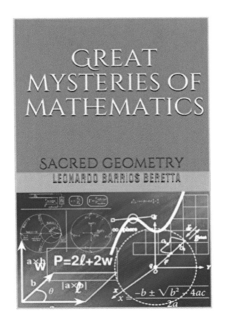 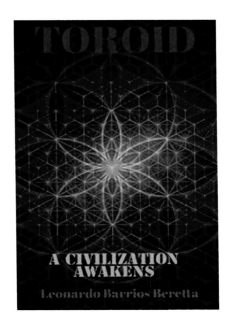

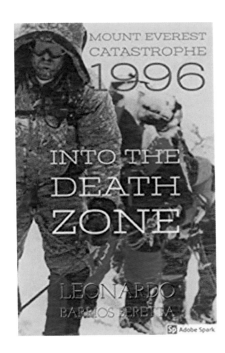 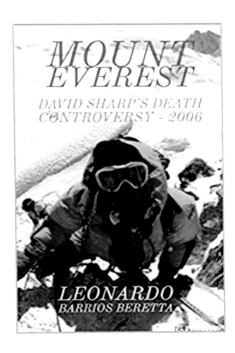

Search for my books in both Spanish and English version...

Made in the USA
Middletown, DE
06 September 2023

38111941R00029